CW0496365

THE REED FIELD GUIDE TO
Common New Zealand
Shorebirds

THE REED FIELD GUIDE TO
Common New Zealand
Shorebirds

David G. Medway

REED

Front cover: Black-phase Variable Oystercatcher.
Back cover: Wrybill in non-breeding plumage (top); juvenile Red-billed Gull passing into immature stage.
Half-title page: Red-necked Stint at Manawatu Estuary.
Title page: Pied Oystercatcher and Caspian Tern at Manawatu Estuary.
Contents page: Department of Conservation sign drawing attention to nesting Northern New Zealand Dotterels and Variable Oystercatchers at Pungaereere Stream.

·Published by Reed Books, a division of Reed Publishing (NZ) Ltd, 39 Rawene Road, Birkenhead, Auckland. Associated companies, branches and representatives throughout the world. Website: www.reed.co.nz

ISBN 0 7900 0738 X
First published 2000

Edited by Carolyn Lagahetau
Cover and design: Sharon Whitaker

Printed in Hong Kong by Toppan Printing

Contents

Sharp-tailed Sandpipers in breeding plumage with Pacific Golden Plovers.

Foreword

Shorebirds make up a significant proportion of the New Zealand bird fauna. The best known of these are the gulls, the terns and perhaps the oystercatchers. Godwits and most of the lesser-known visiting shorebirds have migratory behaviour, nesting in the arctic and travelling thousands of kilometres to spend time here in the southern summer. This annual pattern has fascinated people throughout time.

Shorebirds frequent coastal habitats, which have their own special appeal. The sprawling mudflats of some of the tidal estuaries, the drifting sand dunes of Farewell Spit, and the vast expanse of Lake Ellesmere are a few examples. Several are described, and hopefully this will inspire people to visit them as most are readily accessible.

This book gives a detailed introduction to the more common shorebirds, and attention is drawn to some of the less common species in the hope that this may stimulate readers to look more closely at the birds they see. To help develop this interest, several bird study groups are mentioned and some of their relevant activities are briefly outlined. Many will find this an absorbing hobby.

Brian D. Bell
QSM, Fellow Royal Australasian Ornithologists' Union,
Fellow Ornithological Society of New Zealand
7 April 2000

Introduction

Shorebirds belong to the avian Order *Charadriiformes* which includes waders, gulls and terns. Only the common waders, gulls and terns, being those species which can regularly be seen at one or more of the outstanding or significant New Zealand localities detailed in this book, are fully described and illustrated here. The common New Zealand waders are the Pied Oystercatcher, Variable Oystercatcher, Pied Stilt, Northern New Zealand Dotterel, Banded Dotterel, Wrybill Plover, and the Spur-winged Plover. The common arctic waders are those which occur in this country in the following order of abundance — the Bar-tailed Godwit, Lesser Knot, Turnstone, Pacific Golden Plover, Red-necked Stint, Whimbrel, Curlew Sandpiper, Sharp-tailed Sandpiper, and the Eastern Curlew. The common gulls, all of which are resident in New Zealand, are the Southern Black-backed Gull, Red-billed Gull and the Black-billed Gull. Three of the four terns resident in New Zealand are also common. They are the Black-fronted Tern, Caspian Tern and the White-fronted Tern.

The narrative given in this book relative to each of these species includes detailed sections on their description, population and behaviour. The population sections include a considerable amount of new information, particularly in relation to the numbers and preferred localities of the common arctic waders that spend the southern summer in this country. The species are described in the order set out in the *Checklist of the Birds of New Zealand* (Turbott, 1990), except for the common arctic waders which are described in the order of abundance in which they occur in New Zealand. A brief account of other shorebirds that are resident in New Zealand, or have been recorded in this country, is also included in this book with illustrations of a selection of them.

The two main islands of New Zealand have a number of coastal wetlands, particularly estuaries, which in the New Zealand context are of outstanding or significant importance as habitats for shorebirds. Some of them are described and illustrated in this book. Those chosen as examples of outstanding localities are Kaipara and Manukau harbours and the Firth of Thames in the North Island, and Farewell Spit and Lake Ellesmere in the South Island. Those chosen as examples of significant localities are Manawatu Estuary in the North Island, and Avon-Heathcote Estuary in the South Island. Reasons are given in support of the outstanding or significant status of those seven localities as habitats for shorebirds.

I am grateful to Gwenda Pulham for her guidance during visits to Kaipara and Manukau harbours; to Willie Cook and Rob Schuckard for enabling me to visit Farewell Spit with them; and to Colin Hill, Nick Allen and Ron Nilsson for their hospitality and assistance during visits to Avon-Heathcote Estuary and Lake Ellesmere. I am also grateful to my wife Carole for her understanding while I continue to pursue my interests in ornithology.

David G. Medway
New Plymouth
25 January 2000

New Zealand shorebird localities

Introduction

Wetland ecosystems are characteristic of New Zealand. They are found throughout the length and breadth of the country. Nevertheless, it has been estimated that only about 10 percent of the wetlands present in New Zealand at the time of first human occupation now remain. Many wetland areas have been lost altogether and others have been considerably changed, and continue to be changed, by a variety of human activities. During the more recent period of human occupation, these activities have included the reclamation of estuaries and lagoons and the draining of swamp land; the extension of settlements on to wetlands; the extraction of sand and gravel from rivermouths and riverbeds; and hydroelectric development. In addition, the value of many remaining wetland habitats has been adversely affected by the encroachment into them of a wide range of introduced plants and animals.

Waders, gulls and terns are important, abundant and very visible inhabitants of coastal areas, particularly estuarine wetlands. The waders, the vast majority of which are migratory, are especially dependent upon healthy and safe estuarine habitats. Unfortunately coastal areas, including estuarine wetlands and particularly those near larger centres of population, face considerable and increasing pressure from a variety of human activities including reclamation, effluent disposal and recreation. Many estuaries are now surrounded by urban development, leaving only intertidal areas relatively free of direct encroachment. However, even those intertidal areas are threatened by such factors as the downstream effects of land uses within catchment areas.

Despite these problems, the two main islands of New Zealand still have a number of coastal wetlands, particularly estuaries, which in the New Zealand context are of *outstanding* or *significant* importance as habitats for shorebirds. Only an arbitrarily chosen selection of some of those outstanding and significant localities is able to be individually described and illustrated here. Those chosen as examples of outstanding localities are Kaipara and Manukau harbours and the Firth of Thames in the North Island, and Farewell Spit and Lake Ellesmere in the South Island. Those chosen as examples of significant localities are Manawatu Estuary in the North Island, and Avon-Heathcote Estuary in the South Island.

Whangarei

Kaipara Harbour

Auckland

Manukau Harbour

Firth of Thames

Hamilton

Napier

Wanganui

Manawatu Estuary

Palmerston North

Farewell Spit

Wellington

Westport

Nelson

Avon-Heathcote Estuary

Christchurch

Lake Ellesmere

Dunedin

Invercargill

Stewart Island

Outstanding shorebird localities

Kaipara Harbour

Kaipara Harbour lies north-west of Auckland City. It is New Zealand's largest enclosed harbour and estuarine system, with an area of about 94,700 hectares and over 800 kilometres of coastline. There are a number of small islands around its margins that are connected to the mainland by mudflats or very shallow water at low tide. Kaipara Harbour has a wide range of habitats including intertidal mudflats, sandflats, saltmarshes and saltmeadows. Sand and mudflats are extensive at low tide. The harbour supports a rich and diverse estuarine flora and fauna including a number of threatened species and species that are scarce or local elsewhere in New Zealand. Its extensive mudflats provide a rich food source for up to 50,000 waterfowl, including large numbers of both arctic and New Zealand waders.

Big Sand Island at Tapora, which is a Wildlife Management Reserve, is one of several principal high-tide wader roosts in Kaipara Harbour, and one of the most important shorebird sites in New Zealand. It is one of the least disturbed parts of Kaipara Harbour, and has the largest suitable area within the harbour for roosting waders. Overall, the island probably

supports the largest variety, but not the greatest number, of migratory waders of any site in Kaipara Harbour. Nevertheless, at appropriate times of the year, it has impressive numbers of Bar-tailed Godwits, Lesser Knots and Turnstones, with lesser numbers of a variety of other species. The second Dunlin recorded in New Zealand was seen at Big Sand Island in 1974. The island is also a favoured locality for a post-breeding flock of Northern New Zealand Dotterels.

National wader counts between 1983 and 1994 confirm that Kaipara Harbour is an outstanding shorebird locality for a combination of the following reasons:

• Kaipara Harbour over that period was, along with the Firth of Thames, second only to Manukau Harbour as the locality most favoured by Pied Oystercatchers in winter. Each winter it averaged 13,554, or 17 percent of the New Zealand average winter count, with a highest census count of 21,730.

• It was one of the three localities most favoured by Pied Stilts in winter, and one of their three most favoured North Island localities in summer. Each winter it averaged 2651, or 15 percent of the New Zealand average winter count, with a highest census count of 4679. Each summer it averaged 632, or 10 percent of the New Zealand average summer count, with a highest census count of 1187.

• It was one of the three localities most favoured by Bar-tailed Godwits in both summer and winter. Each summer it averaged 10,381, or 12 percent of the New Zealand average summer count, with a highest census count of 14,507. Each winter it averaged 1173, or 10 percent of the New Zealand average winter count, with a highest census count of 2356.

• Kaipara Harbour over that period was, after Manukau Harbour and Farewell Spit, the locality most favoured by Lesser Knots in summer. Each summer it averaged 7846, or 15 percent of the New Zealand average summer count, with a highest census count of 16,910.

• It was one of the localities most favoured by Turnstones in summer. Each summer it averaged 423, or 10 percent of the New Zealand average summer count, with a highest census count of 618.

• It was one of the localities most favoured by Pacific Golden Plovers in summer. Each summer it averaged 49, or 10 percent of the New Zealand average summer count, with a highest census count of 90.

• It was one of the localities most favoured by Whimbrels in summer. Each summer it averaged 16, or 18 percent of the New Zealand average summer count, with a highest census count of 39.

Kaipara Harbour is also a very important shorebird locality because:

• Most, if not all, Fairy Terns in New Zealand spend the autumn and early winter around the Kaipara Harbour, particularly in the Waikiri Creek area and at Papakanui Spit, which is one of only three known nesting sites of this critically endangered bird.

- Surveys undertaken by members of the Ornithological Society show that Papakanui Spit was second only to the Waitaki River in Otago as the most favoured nesting site in New Zealand for White-fronted Terns during the 1995–96 to 1997–98 breeding seasons.

Manukau Harbour

Manukau Harbour lies immediately to the south and west of Auckland City. It is the second largest harbour in New Zealand with an area of about 34,000 hectares. Approximately 18,000 hectares of the harbour bed is exposed at low tide, revealing a range of intertidal habitats. Most of the harbour catchment is highly modified, but it still retains very significant natural values and is extremely important for arctic and New Zealand waders and other shorebirds. The first Large Sand Dotterel (1943), the first Marsh Sandpiper (1959), the first Dunlin (1969), and the first Baird's Sandpiper (1970) recorded in New Zealand were observed at Manukau Harbour. The one and only Upland Sandpiper so far recorded in New Zealand was seen on the Karaka shellbanks in 1967.

The Karaka shellbanks are one of several principal high-tide wader roosts in the harbour, and one of the most important shorebird sites in New Zealand. However, land access to the shellbanks is through private property and permission to cross must be obtained from the relevant landowners. Many uncommon waders have been recorded on the Karaka shellbanks over the years. The many arctic waders regularly recorded there in summer include very impressive numbers of Bar-tailed Godwits

and Lesser Knots, with lesser numbers of other species such as Turnstones, Curlew Sandpipers, and Red-necked Stints. Among New Zealand waders regularly seen at various times of the year are Northern New Zealand Dotterels, Wrybill Plovers, and very impressive numbers of Pied Oystercatchers.

National wader counts between 1983 and 1994 confirm that Manukau Harbour is an outstanding shorebird locality for a combination of the following reasons:

• Manukau Harbour over that period supported more Pied Oystercatchers in winter than any other locality. Each winter it averaged 25,707, or 32 percent of the New Zealand average winter count, with a highest census count of 31,976.

• It supported more Bar-tailed Godwits in both summer and winter than any other locality. Each summer it averaged 15,534, or 19 percent of the New Zealand average summer count, with a highest census count of 22,571. Each winter it averaged 3110, or 26 percent of the New Zealand average winter count, with a highest census count of 5992.

• More Lesser Knots were supported in both summer and winter than any other locality. Each summer it averaged 16,083, or 31 percent of the New Zealand average summer count, with a highest census count of 22,433. Each winter it averaged 3394, or 64 percent of the New Zealand average winter count, with a highest census count of 5350.

- It supported more Eastern Curlews in summer than any other locality. Each summer it averaged 9, or 31 percent of the New Zealand average summer count, with a highest census count of 19.
- It was equal with the Firth of Thames as the locality most favoured by Pied Stilts in winter, and they were also their most favoured North Island localities in summer. Each winter it averaged 3348, or 19 percent of the New Zealand average winter count, with a highest census count of 4826. Each summer it averaged 779, or 12 percent of the New Zealand average summer count, with a highest census count of 1169.
- It was second only to the Firth of Thames as the locality most favoured by Wrybill Plovers in winter. Each winter it averaged 1171, or 32 percent of the New Zealand average winter count, with a highest census count of 1391.

- Manukau Harbour was one of the localities most favoured by Turnstones in summer. Each summer it averaged 427, or 10 percent of the New Zealand average summer count, with a highest census count of 803.
- It was one of the localities most favoured by Pacific Golden Plovers in summer. Each summer it averaged 47, or 10 percent of the New Zealand average summer count, with a highest census count of 80.

Firth of Thames

The Firth of Thames is a large embayment bounded on the east and west by the Coromandel Peninsula and Hunua Ranges respectively, and to the south by the intensively farmed Hauraki Plains. Most of the water of the Firth is shallow, the inner third all being less than 6 metres in depth. The shallow mud and silt tidal flats of the Firth exposed at ebb tide are esti- mated to cover 8500 hectares. This substantial area of productive tidal flats provides a rich feeding ground for shorebirds, particularly the waders. A large area on the western and southern shores of the Firth of Thames, consisting primarily of shallow estuarine water, mudflat and mangrove forest, is listed as a wetland of international importance under the Ramsar Convention. That part of the Firth is also a member of the East Asian–Australasian Shorebird Reserve Network.

Several species of arctic wader were first recorded in New Zealand on the Miranda coast of the Firth of Thames. These include the Grey Plover (1948), Terek Sandpiper (1951–52) and the Eastern Broad-billed Sandpiper (1960). The first certain Asiatic Black-tailed Godwits recog- nised in New Zealand were seen there in 1955–56. The second Asiatic Dowitcher to be recorded in New Zealand was seen associating with Bar- tailed Godwits on the Miranda coast in 1987.

The Miranda coast between Taramaire and the Pukorokoro Stream, where the Miranda Naturalists Trust Shorebird Centre is located, is the most important shorebird area of the Firth of Thames, and one of the most important in New Zealand. The Stilt Ponds, situated at the southern end of the area, are part of the Robert Findlay Wildlife Reserve which is private property subject to a covenant in favour of the Queen Elizabeth II Trust. They are one of the most important shorebird habitats on the Miranda coast. The Stilt Ponds provide food for many waders, particu- larly Pied Stilts, Banded Dotterels and Sharp-tailed and Pectoral Sandpipers. They are also attractive as a high-tide roosting site for many Bar-tailed Godwits and Lesser Knots. Large numbers of Wrybills frequently roost along their margins, and elsewhere along the Miranda coast, including on the shellbanks at the mouth of the Taramaire Stream. A large area of mudflat near the old limeworks site is a favoured roosting place for impressive numbers of Bar-tailed Godwits, Lesser Knots and Pied Oystercatchers. The shellbanks of this region, as well as providing roosting sites for a large number of shorebirds, are the breeding sites of three species worthy of note. Up to 1000 pairs of White-fronted Terns have

nested on the shellbanks at Taramaire. Black-billed Gulls, and a few pairs of the endangered Northern New Zealand Dotterel, also nest on the shell banks of the area.

National wader counts between 1983 and 1994 confirm that the Firth of Thames is an outstanding shorebird locality for a combination of the following reasons:

• The Firth of Thames over that period supported more Wrybill Plovers in winter than any other locality. Each winter it averaged 1958, or 54 percent of the New Zealand average winter count, with a highest census count of 2702.

• It was equal with Manukau Harbour as the locality most favoured by Pied Stilts in winter, and they were also their most favoured North Island localities in summer. Each winter the Firth of Thames averaged 3452, or 19 percent of the New Zealand average winter count, with a highest census count of 5234. Each summer it averaged 711, or 11 percent of the New Zealand average summer count, with a highest census count of 923.

• Along with Kaipara Harbour, the Firth was second only to Manukau Harbour as the locality most favoured by Pied Oystercatchers in winter. Each winter it averaged 12,618, or 16 percent of the New Zealand average winter count, with a highest census count of 17,657.

• Along with Parengarenga Harbour, it was second only to Manukau

Harbour as the locality most favoured by Lesser Knots in winter. Each winter it averaged 506, or 10 percent of the New Zealand average winter count, with a highest census count of 1835.

• It was, along with Parengarenga Harbour, second only to Lake Ellesmere as the locality most favoured by Curlew Sandpipers in summer. Each summer it averaged 15, or 20 percent of the New Zealand average summer count, with a highest census count of 36.

• The Firth was one of the localities most favoured by Sharp-tailed Sandpipers in summer. Each summer it averaged 13, or 19 percent of the New Zealand average summer count, with a highest census count of 40.

• It was one of the localities most favoured by Whimbrels in summer. Each summer it averaged 19, or 21 percent of the New Zealand average summer count, with a highest census count of 47.

Farewell Spit

Farewell Spit is a nature reserve with public entry by permit only. Tourist traffic to the lighthouse near the tip of the Spit is strictly controlled. The Spit has been a protected area since 1938 in recognition of its outstanding natural values. At that time, almost all of the Spit above high-tide level became a Flora and Fauna Reserve, and the area uncovered at low tide became a Sanctuary for the Preservation of Wildlife. Farewell Spit is listed as a wetland of international importance under the Ramsar Convention. It is also a member of the East Asian–Australasian Shorebird Reserve Network.

Farewell Spit is situated at the northern extremity of Golden Bay and the north-western extremity of the South Island, 38 km from the town of Takaka. It is a recurved spit approximately 25 km long and some 11,388 hectares in area, comprising a land area of about 1961 hectares and an

intertidal zone of about 9427 hectares. Its northern side is exposed to the Tasman Sea but the southern side has a vast area of sandflats, much of it carpeted with saltmarsh, which is covered by the tide for varying periods each day. These tidal flats provide a favoured feeding habitat for a very large number of waders. Following a visit to the Spit in January 1921, one New Zealand ornithologist considered that, 'The bay inside the spit … is one of the great feeding-grounds for the migratory plover in New Zealand.' Another concluded, after a visit in October 1946, that, 'Probably no more prolific feeding ground for waders exists in any other part of New Zealand.'

Mullet Creek catchment, which sometimes becomes an extensive lagoon at high tide, can be one of the main wader roosts on Farewell Spit. Immediately west of Mullet Channel is a saltmarsh flat, only partially covered at high tide, which is a roost for Banded Dotterels, Turnstones, Pacific Golden Plovers, Sharp-tailed Sandpipers and Red-necked Stints. The rest of the catchment is largely bare sand, which is sometimes used as a roost by considerable numbers of Bar-tailed Godwits and Lesser Knots, and a few Eastern Curlews, Whimbrels and other species.

The first New Zealand records of the Mongolian Dotterel and the Western Sandpiper were at Farewell Spit in 1961 and 1964 respectively.

Outstanding shorebird localities

National wader counts between 1983 and 1994 confirm that the Spit is an outstanding shorebird locality for a combination of the following reasons:

• Farewell Spit over that period supported more Banded Dotterels in winter than any other locality. Each winter it averaged 1030, or 13 percent of the New Zealand average winter count, with a highest census count of 1442.

• Along with Parengarenga Harbour, it was the locality most favoured by Turnstones in both summer and winter. Each summer it averaged 846, or 20 percent of the New Zealand average summer count, with a highest census count of 1792. Each winter it averaged 176, or 30 percent of the New Zealand average winter count, with a highest census count of 376.

• It was the South Island locality most favoured by Pied Oystercatchers in winter. Each winter it averaged 7443, or 9 percent of the New Zealand average winter count, with a highest census count of 10,883.

• It was second only to Manukau Harbour as the locality most favoured by Bar-tailed Godwits in both summer and winter. Each summer it averaged 13,557, or 16 percent of the New Zealand average summer count, with a highest census count of 17,181. Each winter it averaged 2626, or 22 percent of the New Zealand average winter count, with a highest census count of 4267.

• It was second only to Manukau Harbour as the locality most favoured by Lesser Knots in summer. Each summer it averaged 15,538, or 30 percent of the New Zealand average summer count, with a highest census count of 24,227.

• Farewell Spit was second only to Manukau Harbour as the locality most favoured by Eastern Curlews in summer. Each summer it averaged 8, or 28 percent of the New Zealand average summer count, with a highest census count of 13.

• It was one of the localities most favoured by Whimbrels in summer. Each summer it averaged 15, or 17 percent of the New Zealand average summer count, with a highest census count of 29.

Lake Ellesmere

Lake Ellesmere (Waihora) is a large, shallow coastal lagoon on the east coast of the South Island about 20 km south of Christchurch City. It is the fifth largest lake in New Zealand with an area of some 20,000 hectares, and about 58 km of shoreline, but this varies significantly depending on water level. The average depth of the lake ranges from 2.5 to 4.5 metres, but water levels continually change throughout the year because of seasonal fluctuations in rainfall, catchment inputs and evaporation rates, wind direction and strength, and because the lake is opened to the sea two or three times a year to lower water levels increased by periodic flooding of surrounding farmland. Lake Ellesmere is the largest brackish lagoon habitat in New Zealand, a habitat type which in itself is uncommon in this country. The Ellesmere wetland is particularly noteworthy for its rushlands and very extensive saltmarsh plant communities, particularly at Greenpark Sands, that are the dominant vegetation types around the shoreline of the lake.

Outstanding shorebird localities

Lake Ellesmere has an impressive list of first records of various species of arctic waders in this country including the Sharp-tailed Sandpiper (1871), Little Whimbrel (1900), Hudsonian Godwit (1902), Red-necked Stint (1902), Curlew Sandpiper (1903), Pectoral Sandpiper (1903), Sanderling (1917), Red-necked Phalarope (1929) and, more recently, the Painted Snipe (1986), Little Stint (1992–93, 1995), Long-toed Stint (1997) and Stilt Sandpiper (1998). Among other waders seen at Lake Ellesmere that have seldom been observed in New Zealand, are the second Grey Phalarope (1925), the second Lesser Yellowlegs (1964), the second and third Ruffs (1984–85), and the second and third Wilson's Phalaropes (1983–84) to be recorded in this country. Thus, all three Phalarope species have been seen at Lake Ellesmere, as have all three Stint species so far recorded in this country.

Lake Ellesmere has outstanding national and international values for wildlife. It is listed as a wetland of international importance under the Ramsar Convention, and it is the most important freshwater wetland in New Zealand for waders.

National wader counts between 1983 and 1994 confirm that Lake Ellesmere is an outstanding shorebird locality for a combination of the following reasons:

- Lake Ellesmere over that period supported more Pied Stilts in summer than any other locality. Each summer it averaged 1110, or 17 percent of the New Zealand average summer count, with a highest census count of 1817.
- It supported more Curlew Sandpipers in summer than any other locality. Each summer it averaged 30, or 40 percent of the New Zealand average summer count, with a highest census count of 59.
- It supported more Red-necked Stints in summer than any other locality. Each summer it averaged 68, or 43 percent of the New Zealand average summer count, with a highest census count of 125.
- It was one of the localities most favoured by Banded Dotterels in winter. Each winter it averaged 887, or 11 percent of the New Zealand average winter count, with a highest census count of 2502.
- Lake Ellesmere over that period was one of the localities most favoured by Sharp-tailed Sandpipers in summer. Each summer it averaged 11, or 16 percent of the New Zealand average summer count, with a highest census count of 48.
- It was the South Island locality most favoured by Pacific Golden Plovers in summer. Each summer it averaged 39, or 8 percent of the New Zealand average summer count, with a highest census count of 128.

Significant shorebird localities

Manawatu Estuary

The Manawatu River estuary is situated at Foxton Beach about 30 minutes' drive from Palmerston North. Because of its central location and ease of access, the estuary regularly attracts members of the Ornithological Society and others from nearby Palmerston North and other parts of the Manawatu, and from the Taranaki, Hawke's Bay, Wairarapa and Wellington regions. The estuary is a point of interest on the Horowhenua Heritage Trails. Information boards on site have photographs of a small selection of birds to be seen there, together with some information about the birds of the estuary, their origins and feeding habits.

Manawatu Estuary is about 200 hectares in area. Although situated in close proximity to an accessible large rivermouth and permanently occupied residential development, with all their associated human activities, it nonetheless retains a high degree of naturalness and diversity. The main freshwater inflow is the Manawatu River, which drains a large catchment area. The wetland itself is of great importance in supporting aquatic and terrestrial food chains. There are extensive areas of exposed mudflats that obviously provide productive feeding habitat for a collectively quite significant number and variety of waders and other shorebirds. In winter, the estuary also supports large numbers of several species of waterfowl and a sizeable population of Royal Spoonbills. The estuary is particularly notable from a birdwatching point of view because of the ease with which its northern side can be accessed, and the closeness to which waders in particular can usually be approached at their principal high-tide roost on that side of the river. Signs have been erected in strategic places in an attempt to deter people with dogs, bikes and other vehicles from that area. At least one local ornithologist, perhaps motivated in part by parochialism, considers Manawatu Estuary to be the best birding site in New Zealand.

Manawatu Estuary is a significant shorebird locality for a combination of the following reasons:

• It is the most important estuarine habitat for shorebirds in the lower North Island.

• Although small in size compared to many other shorebird localities, it supports as great a variety of shorebird species as do more extensive localities where the greatest numbers of shorebirds congregate.

• The estuary is used each year by several species of arctic wader, particularly Bar-tailed Godwits, Lesser Knots, Pacific Golden Plovers, Sharp-tailed Sandpipers, Curlew Sandpipers, and Red-necked Stints. Other arctic waders which are less common include Turnstones, Eastern Curlews, Whimbrels, Terek Sandpipers, Pectoral Sandpipers, and Grey-tailed Tattlers. The first New Zealand records of Great Knot (1967) and Wilson's Phalarope (1983) are from Manawatu Estuary. Other arctic waders seen there, which are uncommon or rare in New Zealand, have been Eastern Broad-billed Sandpipers, and single Hudsonian Godwits and Baird's Sandpipers.

• It is regularly used by New Zealand shorebirds of several species including Pied Stilts, Spur-winged Plovers, Wrybill Plovers, Banded Dotterels, Variable and Pied Oystercatchers, Black-backed Gulls, Red-billed and Black-billed Gulls, and Caspian and White-fronted Terns. Less frequent visitors include Northern New Zealand Dotterels and Black-fronted Terns.

• Manawatu Estuary periodically hosts White-winged Black Terns and Eastern Little Terns. Eastern Common and Arctic Terns have been more unusual visitors. The first and only Red-kneed Dotterel so far recorded in New Zealand was seen at the estuary in 1976.

Avon-Heathcote Estuary

Avon-Heathcote Estuary is only 12 km east of the centre of Christchurch City. With an area of about 880 hectares, it is the largest semi-enclosed estuary in Canterbury. It is one of New Zealand's most important coastal wetlands, despite being almost totally surrounded by residential housing suburbs to the east of Christchurch City and being the focus of regular human activity. The estuary is protected from the Pacific Ocean by the South Brighton Spit. Both the Avon and the Heathcote rivers flow into the head of the estuary from different sides. Associated with the estuary, on its western side, are the Bromley Oxidation Ponds which comprise Te Huinga Manu Wildlife Reserve. The estuary has an average depth of only 1.4 metres at high tide. Extensive areas of mudflats, which are exposed with the receding tide, obviously provide productive feeding habitat for a quite significant number and variety of waders and other shorebirds. In winter, the estuary and oxidation ponds also support large numbers of several species of waterfowl. The plant associations bordering undeveloped sections of the estuary shoreline are dominated by saltmarsh communities.

Avon-Heathcote Estuary is a significant shorebird locality for a combination of the following reasons:

•   The estuary provides habitat for the largest concentrations of New Zealand and arctic waders and other shorebirds on the east coast of the

South Island. It has been estimated that at various times of the year the bird population of the estuary and associated oxidation ponds may include up to 8000 waders, 10,000 gulls and 1800 terns.

• It is a significant summer habitat for Bar-tailed Godwits. Counts show that on average between 1800 and 2200 spent from November to February at the estuary during the period 1987–90. It is also a significant habitat for post-breeding Pied Oystercatchers, with between 3000 and 5000 present there at various times from January to July during the same period. Uncommon arctic waders such as Asiatic Black-tailed and Hudsonian Godwits are occasional visitors. The first Asiatic Dowitcher recorded in New Zealand was seen at the estuary in 1985.

• It hosts an impressive number of gulls over the year. Red-billed Gulls are abundant on the estuary all year, with flocks of over 5000 birds in autumn and winter. Black-billed Gulls are absent from the estuary during part of the year while they breed elsewhere, but most return in February, with winter numbers reaching about 2000 birds. Between 1000 and 2000 Black-backed Gulls are usually present, with higher numbers in winter.

• The estuary regularly supports three species of tern. Flocks of up to 1500 White-fronted Terns have been recorded there, but they are usually much less common. There are normally between 20 and 50 Black-fronted Terns at the estuary during autumn and winter. The estuary is also a very important autumn staging site for some of the Invercargill breeding population of Caspian Terns. The number of that species present at the estuary usually increases to over 100 in March.

Introduction

Several species of wader breed on the two main islands of New Zealand. The **common New Zealand waders** are those seven species that can regularly be seen at one or more of the outstanding and significant shorebird localities described in this book. They are the Pied Oystercatcher, Variable Oystercatcher, Pied Stilt, Northern New Zealand Dotterel, Banded Dotterel, Wrybill Plover, and the Spur-winged Plover.

Several species of wader that breed in the northern hemisphere are present in New Zealand during the southern summer, and others are sometimes present. The **common arctic waders** are those nine species that are in this country each year, although some are present only in small numbers. They can be seen every summer at one or more of the outstanding and significant shorebird localities described in this book. They occur here in the following order of abundance: Bar-tailed Godwits, Lesser Knots, Turnstones, Pacific Golden Plovers, Red-necked Stints, Whimbrels, Curlew Sandpipers, Sharp-tailed Sandpipers, and Eastern Curlews.

Before describing the common New Zealand and arctic waders, it is appropriate to include some pertinent information about the East Asian–Australasian Shorebird Reserve Network, the national wader counts undertaken by the Ornithological Society of New Zealand, the New Zealand Wader Study Group, and the Miranda Naturalists Trust.

The East Asian–Australasian Shorebird Reserve Network

Every year millions of the world's migratory shorebirds travel great distances between their breeding and non-breeding areas. Some species of migratory shorebird weighing as little as 30 grams may migrate up to 30,000 km annually and some species fly more than 6000 km non-stop. Many of the Bar-tailed Godwits that visit New Zealand each summer nest in western Alaska. The route they take from Alaska to New Zealand is still a mystery, but it seems possible that many of them travel the 11,000 km involved directly over the Pacific Ocean! Those migratory shorebirds that make the journey in several stages must undergo several cycles of fat and protein deposition for each sector of the journey. For some species those sectors involve long flights that may last for 36 to 48 hours or more in which stored nutrients are the only source of energy. Failure to deposit enough nutrients would obviously be fatal. Shorebirds have long been known to increase mass before migration.

Most migratory shorebirds make the journey over several weeks. The routes they travel along are called flyways. The East Asian–Australasian Flyway is one of several flyways around the world, but is one of the least understood. It extends from western Alaska and eastern Siberia in the north, down the Asian coast and through South-east Asia to Australasia in the south. An estimated four to six million waders use this flyway in their annual migrations. Each year, between September and March, Australia is host to over two million of those waders. New Zealand is situated at the southern extremity of the flyway. National wader counts carried out by members of the Ornithological Society of New Zealand show that an average total of about 139,000 arctic waders, mostly Bar-tailed Godwits, Lesser Knots and Turnstones, spent each of the summers of 1983–93 in this country.

The Ramsar Convention was signed in 1971. It is an international treaty concerned with the conservation of wetlands of international importance. The broad objectives of the convention are to ensure the wise use and conservation of wetlands of abundant richness in flora and fauna. The convention came into force in New Zealand in 1976. Three of the several New Zealand Ramsar sites are particularly important for waders: the Firth of Thames (in part), Farewell Spit and Lake Ellesmere.

The East Asian–Australasian Shorebird Reserve Network was formally launched during a Ramsar conference in Brisbane in March 1996 in response to the need for international action to protect migratory shorebirds along the East Asian–Australasian Flyway. The East Asian–Australasian Shorebird Reserve Network is an international cooperative effort supported by governments and non-government organisations. It links wetlands that are internationally important for migratory shorebirds and promotes activities for their conservation.

A wetland that meets one of the following criteria is eligible to join the East Asian–Australasian Shorebird Reserve Network:
- it regularly supports more than 20,000 migratory shorebirds, or
- it regularly supports more than 1 percent of the individuals in a population of one species or subspecies of migratory shorebird, or
- it supports appreciable numbers of an endangered or vulnerable population of migratory shorebird.

As at mid-1999, the Network embraced 21 internationally important sites in nine countries. New Zealand currently has two Network sites, both registered in March 1996: the Firth of Thames (in part) and Farewell Spit.

National wader counts

The Ornithological Society of New Zealand initiated the National Wader Count scheme in 1983. Its aims were to determine the numbers and distribution of waders occurring at coastal sites throughout New Zealand; seasonal changes in the distribution and numbers of waders; and annual changes in the numbers of waders. National counts were made each summer during November/early December and each winter during June/early July. The prime objective of the summer counts was to record the number of arctic waders then in this country. The aim of the winter counts was to record the number of New Zealand waders that move to coastal areas following breeding, and to record the number of arctic waders that remained here over winter.

Some 200–250 members of the Ornithological Society assisted with each count. The most extensive series of counts came from Manukau Harbour and the Firth of Thames, where counts began in 1951 and have continued each summer and winter since 1960. Many other localities also received regular attention over the years. They included Kaipara Harbour, Manawatu Estuary, Farewell Spit, Avon-Heathcote Estuary, and Lake Ellesmere. These seven localities, and their importance for waders, are described in this book.

A great deal of information was accumulated about the species involved. For arctic waders the counts produced, among other things:

• reliable estimates of the total numbers of each species in New Zealand

• indications of the year-to-year variations in the numbers of those species

• indications of their breeding success during the previous northern summer

• a better understanding of their distribution within New Zealand.

For New Zealand waders the counts produced:

• minimum population estimates for Wrybill Plovers, Pied Oystercatchers and Pied Stilts

41

• an indication of the proportion of the Banded Dotterel population that remains in New Zealand during winter

• an understanding of the numbers and distribution of all species during winter.

Some of the results of the summer and winter counts carried out from 1983 to 1994 have recently been published in *Notornis*, the quarterly journal of the Ornithological Society of New Zealand. The information thus made available has been of invaluable assistance in the compilation of the locality accounts and many of the species population accounts in this book.

New Zealand Wader Study Group

Banding of arctic waders in New Zealand was started in 1979 at Miranda on the Firth of Thames. It was carried out by a group called the Miranda Banders, which in 1993 changed its name to the New Zealand Wader Study Group to reflect the banding efforts of the group at sites in addition to Miranda. The Wader Study Group operates in association with the Miranda Naturalists' Trust which, among other things, distributes a newsletter that the Group periodically produces.

Members of the Group caught a total of 6975 arctic waders of seven species at seven sites in the Auckland region between 1979 and 1998, mainly using cannon-nets while the birds were at their high-tide roosts on farmland or shellbanks. Almost all of the birds caught were Lesser Knots (5342), Bar-tailed Godwits (1514) or Turnstones (92), with very small numbers of Curlew Sandpipers, Red-necked Stints, Pacific Golden Plovers, and Terek Sandpipers. Thirty of the Lesser Knots and two of the Bar-tailed Godwits had already been banded in Australia. All birds, except for the few which had already been banded overseas, were banded with a metal band bearing a unique number. Since 1991, Lesser Knots, Bar-tailed Godwits, and Turnstones (a total of 1375 birds) have also had a small white plastic leg-flag attached to the tibia. Similar leg-flags, but of different colours, have been applied to birds banded overseas. They are yellow in north-west Australia, orange in Victoria, green in Queensland and blue in Japan. These leg-flags enable observed birds to be instantly identified as having been banded in the localities mentioned.

Of the arctic waders banded or leg-flagged in New Zealand to 1998, up to 21 Lesser Knots, up to 17 Bar-tailed Godwits, and two Turnstones have been recovered or seen in six overseas countries. One Turnstone banded in New Zealand was caught in Australia and then recaptured at its original banding site. As at 1998, up to 135 Lesser Knots, 34 Bar-tailed Godwits, two Turnstones, and two Red-necked Stints bearing Australian leg-flags, as well as two colour-banded Bar-tailed Godwits from Alaska, have been seen in New Zealand. Another two Lesser Knots banded in Australia, and two Bar-tailed Godwits, one banded in Australia and the other in Alaska, have been found dead in New Zealand.

Information so far gained by the Wader Study Group has enabled some deductions to be made as to the likely migration routes taken by the Bar-tailed Godwits, Lesser Knots, and Turnstones that visit New Zealand. However, many more arctic waders will need to be banded in New Zealand, and sightings made of them, before their migration routes can be fully understood.

Between 1987 and 1996, members of the Wader Study Group also banded 2383 Wrybill Plovers on their wintering grounds at two locations near Auckland. As a result, much is being learned about Wrybill Plovers relative to such matters as migration, wintering distribution and movements between sites, longevity, age structure of the population, measurements and body mass, and the timing and duration of feather moult.

Miranda Naturalists Trust

The Miranda Naturalists Trust was incorporated as a charitable trust in 1975 by members of the Auckland Branch of the Ornithological Society of New Zealand to ensure that the Miranda region received the protection and recognition that is warranted for such a scientifically important area. One of the aims of the Trust is to 'establish and maintain an observatory for the study of natural history, especially birds, in the Firth of Thames and adjacent areas'. That facility, known as the Miranda Shorebird Centre and sited on land owned by the Trust on the western side of the Kaiaua–Miranda road, was opened in 1990. For a small charge, accommodation is available to persons who may wish to stay overnight. A manager is resident on the site. Visitors should call at the Shorebird Centre before venturing out onto the shellbanks. Informative displays and publications are available there, as well as useful information about the area, the birds that are currently present, and the best places to see them.

The Trust now has an increasing international membership in excess of 800. For some years it has hosted two open days each year, in October and March, to coincide with the arrival and departure of the arctic waders.

It has commenced residential bird study courses, and has promoted and supported studies that will enable the ecological values of the locality, particularly as they relate to the very significant populations of shorebirds which reside and feed in the area, to be better understood and protected.

Black phase Variable Oystercatcher.

Common New Zealand waders

Pied Oystercatcher *Haematopus ostralegus finschi*
Description Length 46 cm; wingspan 80–86 cm.

The Pied Oystercatcher of New Zealand, commonly known as the South Island Pied Oystercatcher, is generally treated as a subspecies of the widely distributed Pied Oystercatcher. Adults are striking black and white waders, especially in flight. Their head, neck, breast, upper back and tail are black. The lower back, rump, flanks and belly are white. There is a sharp border on the lower breast between their black neck and breast and white belly. A white recess on the 'shoulder' in front of the folded wing is very obvious when the bird is standing. Their upper wings are black with a broad white wingbar, and their underwings are white with dusky tips. In flight, they are black above with conspicuous broad white wingbars and a white wedge on the lower back and rump. They have an orange eye-ring and scarlet iris. Their long, robust, pointed bill is bright orange-red, often with a dusky tip. They have short, coral-pink legs. The sexes are alike in plumage. Males are usually slightly smaller than females with a shorter and redder bill, but these differences are not obvious in the field. Pied Oystercatchers are slightly smaller than Variable Oystercatchers, a difference which is noticeable in the field when both species are together. Immature Pied Oystercatchers have brownish upperparts, a dusky-red bill and pale pink legs.

Population

Pied Oystercatchers are an abundant native bird whose total population has increased very significantly since the species was protected by law in 1940. This protection, coupled with the conversion of tussockland to pasture which increased nesting sites for the species, has probably contributed to their spectacular population increase. The total number approached 49,000 in 1972. It had increased to an estimated 113,000 birds by the winter of 1994.

Nearly all Pied Oystercatchers breed inland in the South Island. Following the completion of breeding, which is usually by late December, most migrate to localities in the northern North Island, principally to

Kaipara and Manukau harbours and the Firth of Thames. Banding studies have shown that individual Pied Oystercatchers usually return year after year to the same wintering locality, and even to the same roost! There is no evidence that juvenile birds accompany their parents to those locations. The huge flocks of Pied Oystercatchers that spend autumn and a good part of winter in those localities reduce considerably in late July to September as birds depart for their breeding grounds. Very little is known about the return routes taken by the multitudes of Pied Oystercatchers that spend the post-breeding months in the northern harbours. In late December/January, a considerable number of birds, in groups both large and small, can be seen flying northwards up the coasts of Taranaki and the Waikato, sometimes just over the land but usually close inshore and only a few metres above the sea. The vast majority appear to fly directly from their breeding grounds to the northern harbours, and directly back again. There is no evidence that a meaningful number stop at any harbours or estuaries on the way.

National wader counts during the winters of 1984–94 show that, although the annual average of 80,619 Pied Oystercatchers recorded during that period was widely distributed throughout the country at that time of year, most birds (an annual average of 51,879 or 64 percent) were present then at just three localities: Manukau Harbour (with an annual average of 25,707 birds), Kaipara Harbour (with an annual average of 13,554 birds), and the Firth of Thames (with an annual average of 12,618 birds). About 55 percent of the annual average of 23,251 birds that spent the winters of those years in the South Island were recorded in the Nelson area.

The increase in the number of Pied Oystercatchers that have wintered during the last few decades at Manukau Harbour and the Firth of Thames combined is remarkable. Probably fewer than 1000 birds wintered there in 1941. That number had increased to annual averages of 6568 during the winters of 1960–69; 21,733 during the winters of 1970–79; 33,704 during the winters of 1980–89; and 47,281 during the winters of 1990–98! The percentage of the total New Zealand population of Pied Oystercatchers that wintered at Manukau Harbour and the Firth of Thames combined

also increased from 34.5 percent in 1972 to 47.5 percent for the period from 1984 to 1994.

An annual average of 14,779 Pied Oystercatchers were recorded on estuarine areas throughout the country during each of the summers of 1983–93. At this time of year most adult birds would be at inland breeding sites. Those that remained on the coast were mainly subadult birds, most of which were recorded at the main wintering sites, particularly at Manukau Harbour and the Firth of Thames, and in the Nelson area.

Behaviour

Pied Oystercatchers breed between August and January, mainly inland east of the Southern Alps from Marlborough to Southland, and mainly on

Pied Oystercatcher showing white 'mirror' in front of folded wing.

shingle riverbeds and farmland. Established pairs usually reclaim the same nesting territory year after year, even though the male and female seem to spend the winter at different localities. Nesting birds mob aerial predators, and attempt to lead human intruders away from the nest, and particularly from chicks, with aerial attacking, and running with false brooding and feigned wing injury, accompanied by loud and persistent calling. Trampling by stock and farming activities are among the main causes of egg loss for Pied Oystercatchers that nest on farmland in Mid-Canterbury.

Pied Oystercatchers form huge roosting and feeding flocks at their wintering locations. They feed mainly on molluscs, estuarine worms, earthworms, and insect larvae, but other small invertebrates and small fish are taken. They feed in estuaries, on sandy shores, in pastures, in ploughed paddocks and riverbeds by surface picking and deep probing. Vast amounts of food must be consumed by the huge numbers of Pied Oystercatchers that spend from about January to August at Kaipara and Manukau harbours and the Firth of Thames. Studies at Avon-Heathcote Estuary indicated that 4000 Pied Oystercatchers wintering there consumed the amazing total of about 1,472,000 cockles per day!

Pied Oystercatchers roosting with Bar-tailed Godwits and Lesser Knots at Miranda in the Firth of Thames.

Variable Oystercatcher *Haematopus unicolor*
Description Length 47–49 cm; wingspan 83–89 cm.

Variable Oystercatchers are slightly larger than Pied Oystercatchers, a difference that is noticeable in the field when both species are together. Adults have a long, robust, somewhat blunted bill which is bright orange often with a yellowish tip, short, stubby, coral pink legs sometimes with a purple tinge, and an orange eye-ring and scarlet iris. The sexes are alike in plumage. Males are usually slightly smaller than females with a shorter and redder bill, but these differences are not obvious in the field. The plumage of Variable Oystercatchers varies from black to pied with a continuous gradient between. Pied phase birds are similar to Pied Oystercatchers, but they have a larger, heavier bill and a band of mottled feathers between the black and white plumage of the breast. Even in the most pied individuals the white recess on the 'shoulder' in front of the folded wing, which Pied Oystercatchers show so clearly when standing, is seldom present, but there is usually some patchy white feathering in that area. Pied phase birds also have a white wingbar and rump in flight, but their lower back is black or smudgy whereas in Pied Oystercatchers it is white. Intermediate phase birds have variable amounts of white on the

Black phase Variable Oystercatcher.

wingbar, rump and belly that can be very smudgy in appearance. Black phase individuals are entirely black. The relative abundance of the three colour phases varies with latitude. In the northern North Island, about 43 percent are black, in central New Zealand about 85 percent are black, and in the southern South Island and Stewart Island about 94 percent are black. Immature Variable Oystercatchers have brownish upperparts, a dusky-red bill and pale pink legs.

Population

Variable Oystercatchers are a relatively common endemic species. In the North Island they are most abundant along the north-eastern coast from North Cape to Mahia Peninsula, and near Wellington. In the South Island they are common around Tasman and Golden bays, Marlborough Sounds and Fiordland. They are also common on the beaches of Stewart Island and its offshore islands. Their total pop-ulation appears to have increased significantly since they were protect-ed from shooting in 1940. This is especially so in the northern North Island where many nests sited on sandspits are now also protected from mam-

Black phase Variable Oystercatcher.

malian predators and human disturbance. A comprehensive survey of coastal New Zealand resulted in a population estimate of 2000 birds in 1970–71. National wader counts between 1984 and 1994 produced an annual average winter population of only 1393, with a highest winter count of 1849 in 1989. About twice as many were counted in the North Island as in the South Island. However, the wader count total no doubt greatly underestimates the true population size because those counts did not cover Fiordland and southern Stewart Island where many Variable Oystercatchers are known to occur. Other recent survey data from the complete New Zealand coast, including Fiordland and Stewart Island, produced an estimate of 4000 birds, again with about twice as many being in the North Island. This indicates an approximate doubling of the total population since the early 1970s.

Behaviour

Variable Oystercatchers breed from September to February on rocky and sandy coasts of the North and South islands and Stewart Island and its offshore islands. It is not unusual for birds of different colour phases to form a breeding pair. Some pairs of Variable Oystercatchers seem to occupy their territory throughout the year, but others may gather at estuaries in autumn and winter. Some form small winter flocks, sometimes with a few Pied Oystercatchers. The occasional individual Variable Oystercatcher may be seen among large flocks of that species. Nesting birds mob aerial predators, and attempt to lead human intruders away from the nest with close aerial attacking, and running with false brooding and feigned wing injury, accompanied by loud and persistent calling. Adults noisily lead their young away from humans and hide them under or among logs and branches or other debris where they are undetectable except with the closest and most persistent searching. Young birds, which can swim well from a very early age, are occasionally attacked by Black-backed Gulls, which are usually driven off by nearby adult oystercatchers. Immature birds sometimes wander extensively before pairing and settling in one area.

Variable Oystercatchers feed mostly on molluscs, worms and crabs, and also on other small invertebrates and small fish. They feed mainly at estuaries and on sandy shores, and among pools on rocky coastlines, by surface picking and deep probing, but will sometimes venture into coastal fields after heavy rain to catch earthworms and insect larvae.

Pied phase Variable Oystercatcher.

Pied Stilt *Himantopus himantopus leucocephalus*
Description Length 33–37 cm; wingspan 61–73 cm.

The Pied Stilts of New Zealand belong to a subspecies that extends from the Philippines, Indonesia and the Bismarck Archipelago to Australia and New Zealand. They are unmistakable slim black and white waders with very long pinkish-red legs that extend well beyond the tail in flight, and a long, fine, straight, black bill. The sexes are alike. Adult birds have a white head. Their hind neck is black. In some birds a black collar completely encircles the lower neck. They also have a black back, and both the upper and undersides of their wings are black. Their underparts and upper tail are white. Most immature birds have a variable amount of grey on the crown which can extend downwards to include the eyes, but some have almost pure white heads. They do not have the black hind neck of adults. Juveniles have a grey wash on their head and neck, a dark brown back and wings, and dull pink legs.

Adult Pied Stilt.

Population

Pied Stilts are probably relatively recent colonists of New Zealand from Australia. They are now common throughout lowland parts of the country with a population estimated at about 30,000 birds. Pied Stilts nest between July and January. During December to February birds that breed inland move to coastal areas, but those that breed on the coast or in northern parts of New Zealand are usually sedentary throughout the year. Many of those sedentary birds, particularly those on small wetlands in the northern half of the North Island, would not be recorded in national wader counts. Many South Island and southern North Island birds migrate to northern parts of the North Island after the breeding season. Pied Stilts return to their breeding grounds in June–July in lowland places and August–October in inland places.

Pied Stilts are widespread in both main islands during winter, but between 1984 and 1994 about 85 percent of those recorded in national wader counts during that season were in the North Island, the highest numbers being consistently present at Kaipara and Manukau harbours and the Firth of Thames. Lake Wairarapa was also a favoured locality. Each of the Firth of Thames and Manukau Harbour averaged about 3400 Pied Stilts in winter during that period, while the Kaipara Harbour averaged somewhat less at about 2600 birds. Winter counts of Pied Stilts at Manukau Harbour and the Firth of Thames over the period 1960–98 showed little change over time in the number of Pied Stilts present at those localities during that season. Lake Ellesmere was the South Island locality most favoured by them during the winters of 1984–94, with an annual winter average of 548 birds.

During summer counts between 1983 and 1993, only about 37 percent of the number of Pied Stilts counted in winter were recorded. Most birds would have returned to their breeding grounds by the time those counts were made, those remaining being local breeders and immature non-breeders. The majority of Pied Stilts counted during summer in the North Island were again present at Kaipara and Manukau harbours and the Firth of Thames, but numbers were considerably smaller there in summer. Manukau Harbour and the Firth of Thames each averaged about 750 birds, while Kaipara Harbour again averaged somewhat less at about 650 birds. Lake Wairarapa was also a favoured summer locality. Lake Ellesmere was the locality most favoured by Pied Stilts in New Zealand during the summers of 1983–93, with an annual average of 1110 birds throughout that period.

Behaviour

Pied Stilts usually nest in loose colonies of up to 20 pairs but sometimes many more. They nest throughout most of New Zealand close to water in both coastal and inland locations on open areas in swamps, lagoons, damp fields, gravel riverbeds, estuaries and saltmarshes. They strenuously

defend their nest sites from intruders with a variety of distraction displays, including feigning flightlessness accompanied by calls of distress. They are gregarious during all seasons. They form flocks at estuaries and lakes outside the breeding season where they roost together, often in large, compact groups. Pied Stilts are nervous and excitable birds. They yap persistently and noisily when feeding and flying both during the day and at night. They feed at shallow lagoons and ponds, tidal flats, swamps, lake edges, and the lower reaches of rivers. In some coastal localities, large numbers can sometimes be seen feeding on wet pasture, often in the company of Red-billed Gulls and other birds. They feed throughout the day and sometimes at night, but mainly in the early morning and late afternoon. They walk through shallow water pecking at the surface or plunging their heads underneath. They eat mainly water insects, but they also take small molluscs and earthworms.

Northern New Zealand Dotterel *Charadrius obscurus aquilonius*
Description Length 26–28 cm; wingspan 46–50 cm.

New Zealand Dotterels are large plovers with a large head, large dark eyes and robust black bill. Two subspecies are recognised, the breeding populations of which are widely separated. The larger Southern New Zealand Dotterel breeds only on Stewart Island. The smaller Northern New Zealand Dotterel has a restricted distribution and breeds mainly on the coast of northern North Island. Only the Northern New Zealand Dotterel is considered here. Adults are generally in breeding plumage from June to December, but some begin to assume it in April. They have brown upperparts that are finely streaked dark brown with whitish feather edges.

Adult Northern New Zealand Dotterel.

Males have red breasts and bellies, the red varying in extent and intensity from dark chestnut to a pale cinnamon. Females usually show some red, but occasionally breeding pairs show very little reddish colouration on the underparts. They have a white forehead and pale superciliary streak. The bill is black and slightly upturned at the tip, and the legs are olive-grey. The sexes are similar in non-breeding plumage. They have brown upperparts, with paler edges to the feathers, and whitish underparts with a very pale brownish breast band that is often restricted to the shoulder area. Juveniles are like pale adults in breeding plumage. They have fresh white underparts, but with a pale orange wash on the breast and belly. Some may also have a narrow band of grey spotting from the shoulder across the breast. They have flecked grey feathers on the back and head.

Population

Northern New Zealand Dotterels are a threatened endemic subspecies. With one known exception, their whole population breeds from North Cape to Taharoa Beach near Kawhia Harbour in the west, and to Mahia Peninsula in the east. However, in the 1998–99 breeding season, a resident pair of Northern New Zealand Dotterels successfully raised one chick on a sandy beach just south of Cape Egmont. This is the most southern known breeding record on the west coast of the North Island.

It is not difficult to see Northern New Zealand Dotterels at suitable localities within the regions where they nest. Individual birds are occasionally recorded at other North Island localities, particularly on the west coast south of their normal breeding range. Many of these may be juvenile birds, which are known to wander widely. Some adult pairs are entirely sedentary on their breeding grounds. However, after nesting, many Northern New Zealand Dotterels and their progeny travel a short distance from their breeding grounds to form post-breeding flocks at favoured coastal sites. Post-breeding flocks begin to form in January and are at peak numbers in February and March. Birds begin to return to their breeding grounds in late March, and the majority of those that bred away from the flock site have left by the end of April. A significant proportion of the population of Northern New Zealand Dotterels can therefore be surveyed each autumn by counting the number of birds in these post-breeding flocks. Not all flocks are suitable for this purpose because at some sites numbers fluctuate markedly within weeks or even days, probably because the birds have alternative roosts nearby. Counts of post-breeding flocks of Northern New Zealand Dotterels are carried out each autumn by members of the Ornithological Society and others interested in their welfare. Four Auckland localities have given consistently good results over recent years. In the autumn of 1999 they together produced a total of 305 Northern New Zealand Dotterels. Those localities are Mangawhai and Omaha on the Auckland east coast which produced 131 and 62 birds respectively, and Big Sand Island and Papakanui Spit in Kaipara Harbour which produced 51 and 61 birds respectively. The total population of the Northern New Zealand Dotterel is estimated to be about 1400 birds. Thus, in autumn 1999, about 22 percent of the estimated total population of the subspecies was accounted for at those four localities. Autumn counts have shown that Mangawhai is the most important post-breeding flocking site for Northern New Zealand Dotterels.

New Zealand Dotterels are the subject of a recovery plan. A recovery group set up to implement and oversee the plan comprises members from the Department of Conservation, Ornithological Society and Royal Forest and Bird Protection Society. Northern New Zealand Dotterels, like nearly all birds, are most vulnerable when nesting. Their nest sites, particularly those on dune areas and beaches, can be adversely affected by many factors including high spring tides and storms, shifting sand, wandering stock, human development and disturbance, and

predation by both introduced mammals (including stoats which take eggs, chicks and adults, and hedgehogs which take eggs), and other birds (including Black-backed Gulls which take eggs and chicks). Northern New Zealand Dotterel nests are very vulnerable to predation even when mammalian predators are in low numbers. Studies have shown that very few young are being recruited into the general population, and that this low productivity is mainly caused by predation of eggs and chicks, and by human disturbance. Northern New Zealand Dotterels can live a long time, several being known to have survived for more than 20 years. This long life expectancy may be the prime factor responsible for maintaining their total population at over 1000 birds during recent decades. To help them breed successfully, considerable publicity is given to their plight. Some feeding and nesting sites are protected from intrusion by humans and their associated animals such as horses and dogs, predators are trapped around selected nest sites, breeding is monitored, and some chicks have been captive-reared and released back in the wild. The protection of selected nest sites has greatly improved the breeding success of the birds involved. However, this protection is not available everywhere, and in some places nesting habitat is diminishing. At present, Northern New Zealand Dotterels appear to be holding their own, or perhaps even increasing slightly in number as a result of nest protection.

Behaviour

Members of most established pairs of Northern New Zealand Dotterels remain together throughout the year. The pair-bond appears to be continuous. It lasts for a number of consecutive seasons, and possibly for life, in most or all cases. Breeding Northern New Zealand Dotterels are usually on their nesting territories by early August. They have a protracted egg-laying period that extends from late August to mid-January. During this time a pair may nest up to four times if a clutch is lost. Recently, several pairs have been observed to lay again after young from their first clutch had fledged. They usually nest on low flat sites with little or no vegetation such as sandspits, sandy beaches, low sand dunes, shellbanks and stream mouths. Northern New Zealand Dotterels regularly nest close to nesting Variable Oystercatchers, and sometimes near to or even among nesting White-fronted Terns. They display very high breeding-territory fidelity, with nearly all pairs occupying the same or very similar territories from one season to the next. They will readily adopt suitable new nesting sites, whether naturally or artificially created.

Northern New Zealand Dotterels are often quite confiding and approachable, particularly when roosting in a post-breeding flock. They are very territorial during the breeding season. They usually strenuously attempt to divert intruders of all sorts from their nest or chicks by employ-

ing a wide variety of tactics. Birds with eggs will run or fly noisily at an intruder. If the intruder is persistent they feign injury by trailing one wing and one leg, ruffling their feathers, drooping their tail, and making calls of distress. These displays are most intense when eggs are close to hatching. Birds attempt to draw an intruder away from young chicks by running in a low, hunched posture, calling loudly and frantically all the time. Young chicks are very mobile. They often hide among vegetation, or under or among logs and branches or other debris, where they are undetectable except with the closest and most persistent searching.

Northern New Zealand Dotterels run quickly and pause while feeding. They feed mainly along the upper tidal levels of the foreshore where it is sandy or stony, at stream mouths, on sandy beaches and dunes, and on the short turf of seaside pastures with other gregarious plovers such as Banded Dotterels and Wrybills. In at least one locality, they regularly feed on a rocky foreshore in and among pools left by the receding tide. They sometimes tap vigorously with their feet in shallow sea water in order to bring small aquatic creatures to the surface, or along the driftline to flush prey such as sandhoppers. They feed on crustacea, small mollusca, marine organisms including small fish, and various sorts of insects. Some Northern New Zealand Dotterels were recently seen eating the tips of glasswort plants. This is the first record of New Zealand Dotterels eating plant material.

Banded Dotterel *Charadrius bicinctus bicinctus*
Description Length 18–21 cm; wingspan 37–42 cm.

Banded Dotterels are small, plump plovers with a rather slender, dark grey bill, variable greenish-yellow legs, and upright stance. They have a confusing range of plumages according to age, sex and season. Adult birds are easily recognised in breeding plumage, usually from May to January, when both male and female have two distinctive breast-bands. They are the only plover in New Zealand with two breast-bands. The upperparts are brown, and the underparts are white except for a thin band on the lower neck that is brown in females and blackish in males, and a broader band on the breast that is chestnut-coloured in both sexes. The male has bolder and darker bands, but their distinctness varies considerably. The forehead is white, edged above with black in the male, which also has a dark line through the eye. Birds in non-breeding plumage, usually from February to May, are very variable. The male loses its black facial markings. The lower breast-band fades and is often lost. Some individuals retain a distinct upper neck-band, but it usually remains only as a faint incomplete necklace of spots. The plumage of juvenile birds is like that of non-breeding birds except that the whole head is washed yellowish-buff, they usually have dark shoulder patches, and the upperparts are greyish-brown with fawn or off-white edges to the feathers.

Juvenile Banded Dotterel.

Population

Banded Dotterels are an abundant endemic species. Their breeding concentrations are on the shingle riverbeds of Hawke's Bay, Manawatu and the Wairarapa in the southern North Island, and on the braided riverbeds of Marlborough, Canterbury, Otago and Southland in the South Island. Their main breeding stronghold is Canterbury, where an estimated 10,000 pairs nest. The nesting sites of Banded Dotterels on braided riverbeds of the eastern South Island are vulnerable because of such factors as variable river flows, encroachment by introduced plants, introduced mammalian predators, and human recreational use. Predation by introduced mammals, including the loss of eggs to stoats and hedgehogs, appears to be the main cause of failure of Banded Dotterel nests on at least some of those riverbeds. Some of these factors no doubt also affect birds nesting elsewhere in New Zealand. Nevertheless, Banded Dotterels appear to be in a healthy position overall with a total population estimated at about 50,000 birds, most of which are in the South Island.

Banding studies have shown that regional populations of Banded Dotterels have different post-breeding movement patterns which range from sedentary behaviour, through migration within New Zealand, to trans-Tasman migration. Birds breeding at

Male Banded Dotterel in breeding plumage.

coastal sites throughout New Zealand are mainly sedentary. Those breeding inland north of Canterbury mostly migrate within New Zealand, particularly to harbours in the North Island, but there are regionally specific patterns. In southern New Zealand, birds on coastal and inland breeding grounds separated by only short distances may have very different post-breeding movement patterns. The nominate race of the Banded Dotterel is unique among waders in that a large part of the population undertakes an east–west migration. Sight recoveries of birds colour-banded on their breeding grounds have shown that most birds breeding in inland and high-altitude regions from North Canterbury southwards, possibly as many as 30,000, migrate to wintering areas in south-eastern Australia, a minimum distance of 1600 kilometres.

National wader counts from 1984 to 1994 recorded an average of just 7882 Banded Dotterels in the winter of each of those years. This is only about 39 percent of the estimated 20,000 thought to be in New Zealand at that time of year. However, many Banded Dotterels were almost certainly at localities that were not included in the counts. Nevertheless, the counts did show that, of the localities covered, Parengarenga and Manukau harbours in the North Island, and Farewell Spit and Lake Ellesmere in the South Island were those most favoured by the Banded Dotterels that remained in New Zealand after breeding.

Behaviour

From December until July, flocks of up to several hundred Banded Dotterels are distributed around the New Zealand coast where they can be found in a wide range of habitats. These include estuaries, sandy beaches, the mouths of streams, the margins of coastal lakes and ponds, saltmarshes, coastal farmland, ploughed fields, and even the grassed areas of some airports. Some Banded Dotterels return to their breeding grounds as early as mid-July, but most do so between mid-August and early September. They nest throughout New Zealand between then and January in a wide range of habitats, both coastal and inland, including inland shingle riverbeds and river terraces, inland gravel lakeshores,

subalpine herbfields, coastal riverbeds, rivermouths, sandy and gravel beaches, coastal lagoons, estuaries and pastures. Some unusual sites where pairs have nested include stony ground at a coastal petrochemical plant and at disused railway yards.

Banded Dotterels feed in loose flocks or singly, running or walking and stopping at random to peck or probe. They are opportunistic feeders which rely heavily on invertebrates of aquatic and near-aquatic sources, but they have also been recorded eating the berries of prostrate plants. The encroachment of introduced plants such as willow, goose, broom and lupin has seriously degraded the value of some of their feeding habitat on the braided riverbeds of the eastern South Island.

Wrybill Plover *Anarhynchus frontalis*
Description Length 20–21 cm; wingspan 38–43 cm.

Wrybill Plovers (commonly known as Wrybills) are distinctive small, grey and white plovers about the size of a Banded Dotterel, but they have a long, narrow, black bill, which is unique in that the last third turns to the right. Males and females can only be reliably distinguished when they are in breeding plumage. The crown, nape and upperparts of adults in breeding plumage are plain ashy grey. They have a white forehead, which in the male is edged above with a thin, black frontal band. Their underparts are white except for a black band across the upper breast which is wider in the male. In non-breeding plumage males do not have the black frontal band, and in both sexes the breast band is indistinct or absent. In juvenile birds the back feathers are edged with white and the breast band is always absent.

Population
Wrybills are a threatened endemic species. They nest between late August and December, usually close to water on braided riverbeds in Canterbury and inland Otago. The Rakaia River is the most important Wrybill breeding locality with more birds and habitat suitable to them than any other river. Their nests are vulnerable because of such factors as flooding, predation by introduced mammalian predators including stoats, and human recrea-

tional use. Their principal bird enemies are Australasian Harriers and Black-backed Gulls, but flooding is probably the main cause of losses of eggs and chicks. The value of some of their feeding and breeding habitat on braided riverbeds has been seriously degraded by the encroachment of introduced plants such as willow, gorse, broom, and lupin.

Juveniles from early nests head north at the end of December, while most adults leave in mid-January or later. Wrybills begin arriving at their principal wintering grounds in the harbours of Northland, Auckland and South Auckland during late December and January. Adults are usually faithful to their past wintering sites, but a minority have been recorded as using more than one site during a season. The main route they take from their South Island breeding grounds is not known. Most birds appear to migrate directly, with only a few stopping temporarily on the way at such places as Lake Ellesmere and Manawatu Estuary.

National wader counts between 1984 and 1994 produced an average of 3635 Wrybills in the North Island during winter, with an average of only 23 in the South Island. At the time of year when those counts were made, the majority of Wrybills were congregated at the Firth of Thames and Manukau Harbour where the number present each winter during the period referred to averaged 3129 birds, being approximately 86 percent of the average winter population of the species. The Firth of Thames was the more favoured of those two localities with an average of 1958 birds (or 54 percent of the average winter population), while Manukau Harbour averaged 1171 (or 32 percent of the average winter population). Other localities had only very small numbers of Wrybills during winter compared to the Firth of Thames and Manukau Harbour. Those that averaged more than 100 during the winters of 1984–94 were Parengarenga Harbour (137), Whangarei Harbour (136), and Kaipara Harbour (115). There is some reason to believe that the national wader counts may have under-recorded the actual number of Wrybills then present in New Zealand. A nationwide count specifically of Wrybills on 29 May 1994 produced a total of 5111 birds, compared to a total of 4197 birds produced during the national wader counts held slightly later in June/early July of that year. Some 4261, or about 83 percent of the 5111 birds counted on 29 May 1994, were found at the Firth of Thames and Manukau Harbour.

The first adult Wrybills begin to move south at the end of July, with the majority departing in mid-August. Many young birds follow them about a month later, leaving few Wrybills remaining in the North Island. Breeding pairs reappear on Canterbury riverbeds in August. The main route Wrybills take back to their South Island breeding grounds is not known. The 800-plus birds that were at Manawatu Estuary for a very short time in mid-October 1986 is a number quite unknown at that locality, before or since. Many of the birds involved may have been young individuals, no doubt on passage to the South Island, which had stopped there briefly for some unknown reason.

Behaviour

Wrybills roost in dense flocks, usually apart from most other waders, but some small waders like Curlew, Terek and Sharp-tailed Sandpipers and Red-necked Stints will often be found resting among them. They frequently roost on sparsely vegetated seaside flats among low growth like glasswort. Wrybills are the most confiding and approachable of waders when at rest on their wintering grounds. They roost quietly together. Almost all rest on one leg and when disturbed they prefer to hop away on it rather than run on both. Large flocks of Wrybills can frequently be seen on their wintering grounds, particularly before southward migration begins, performing swift and spectacular aerial displays in which hundreds of birds swirl as one in the air with the sun catching their white undersides as they turn. Often on landing they all run swiftly for a short distance before stopping. They are partial to bathing together in shallow pools left by the receded tide.

Male Wrybill in breeding plumage.

Female Wrybill in breeding plumage.

Wrybills are opportunistic feeders that rely mainly on invertebrates of aquatic and near-aquatic sources. They feed primarily within or near shallow water edges. They commonly feed by pecking insects off the water surface, or by tilting their head to the left followed by clockwise movements of the bill under stones in shallow pools. Studies on the Rakaia River show that the diet of Wrybills nesting there consists mostly of mayfly larvae, bugs, beetles and flies. On the tidal mudflats of their wintering grounds, Wrybills feed on small crustaceans which they prefer to catch in silty mud with a surface film of water. They often begin to feed as soon as the tide starts to ebb while most other waders continue to roost. Redbilled Gulls have been observed robbing Wrybills of food.

Between 1987 and 1996, about 2400 Wrybills were banded on their wintering grounds, and more were colour-banded on their breeding grounds. Colour-banded Wrybills can often be seen among roosting flocks. As a result of these continuing activities, a great deal is being learned about Wrybills relative to such matters as migration, wintering distribution and movements between sites, longevity, age structure of the population, measurements and body mass, and the timing and duration of feather moult.

Spur-winged Plover *Vanellus miles novaehollandiae*
Description Length 30–37 cm; wingspan 75–85 cm.

The Spur-winged Plovers of New Zealand belong to a subspecies that is also found from mid-Queensland to south-eastern Australia and Tasmania. They are conspicuous large plovers. They are broad-winged and strikingly patterned, and quite unlike any other bird in New Zealand. Adults have a black crown and hindneck, and a broad, black stripe passing downwards on each side of the body from the hindneck to the sides of the breast in front of the folded wings. They have a brown back. Most of the upperwing coverts are also brown, but the primary feathers, primary coverts and most secondary feathers are black, giving the upperwing a brown and black patterning which is distinctive in flight. The rump is white, as is the upper tail, which has a conspicuous black band near the narrow white tip. Their underparts are white. The underwings are also white with a broad, dark trailing edge which is obvious in flight. They have a yellow bill, large, bright yellow wattles, a bright yellow iris and black pupil, and long, reddish legs and feet. Their long and sharp bony wingspurs, which are yellow and black-tipped, are often concealed by the breast feathers. They are exposed most often when a bird is involved in displaying, or is settling on to the nest, and when incubating. The sexes

Spur-winged Plover in aggressive mood showing spurs on wings.

are alike, and there are no seasonal differences. Juveniles have dull yellow wattles which are initially small but grow quickly, their bill is a dull yellow with a dusky tip, the feathers on their upperparts and upperwings are narrowly edged black and buff, and their legs and feet are dark grey. Immature birds are very similar to adults, except some retain a few mottled upperwing coverts of the juvenile stage.

Population

Spur-winged Plovers have prospered in New Zealand since they recently introduced themselves from Australia. A pair bred in New Zealand for the first known time at Invercargill Airport about 1932. During the 1950s they expanded to inland Southland. From the late 1960s to the 1980s they spread to Stewart Island and to the rest of the South Island. They are now abundant in Southland, coastal Otago and Canterbury, and common in all other suitable areas. The first breeding records in the North Island were in the early 1970s. The species is now abundant in coastal Manawatu and the southern Wairarapa, and is either already common or very common, or is rapidly becoming so, throughout the rest of the North Island. It is not possible to estimate the New Zealand population of Spur-winged Plovers from the twice-yearly national wader counts because large numbers occur away from the places where those counts are made. The New Zealand population of the species is not known, but it probably numbers in the tens of thousands, and is undoubtedly still growing.

Behaviour

Spur-winged Plovers fly buoyantly with slow, deliberate beats of their rounded wings. They sometimes tumble suddenly in the air. Extreme noisiness is probably their most noticeable feature. They have a pene-trating call, variously described as a 'noisy grating rattle' or a 'loud stac-cato rattle', which is given by day and sometimes by night, both when on the ground and in the air. This unmistakable call frequently confirms the presence of the species although the birds responsible for it may never be seen. Spur-winged Plovers are very wary and alert. A group will quickly take noisy flight on the approach of a human intruder even while the per-son is still some distance away. This can be particularly frustrating at wader roosts because the departing Spur-winged Plovers will almost invariably cause any other resting birds to become airborne with them. They have a variety of distraction displays, and are aggressive, particularly

in defence of their nest and chicks. They can often be seen attacking Australasian Harriers and Australian Magpies as they fly over, and they have been observed in pursuit of Black-backed Gulls. On the other hand, one solitary Spur-winged Plover standing in a coastal saltmarsh paddock was persistently harassed from above by a determined White-winged Black Tern which twice succeeded in making the much larger bird take briefly to the air.

Spur-winged Plovers have a long breeding season, from June to December, with a peak of egg laying in August. They nest in a variety of grassland and shingle riverbed habitats. Spur-winged Plovers are often gregarious when not breeding. Flocks of many different sizes may congregate at some airports, and on farmlands, wetlands with short vegetation, and at estuaries both large and small. Groups of up to 600 birds have been recorded at Lake Wairarapa, but their flocking activities appear to be local, and no long-distance movements have yet been noted in New Zealand. The extensive open pastures and arable farmlands to be found in New Zealand clearly provide much habitat suited to the species. In the South Island they can sometimes be seen feeding in the same paddock as Black-billed Gulls, Black-fronted Terns, Banded Dotterels, and Pied Oystercatchers. Their diet on short-grassed pasture and ploughed paddocks is mainly earthworms and insects and their larvae, as well as seeds and leaves. Their feeding preferences in these habitats are no doubt beneficial to farmers because the insects and larvae that Spur-winged Plovers eat include pests such as grass grubs and porina moth larvae. In tidal and other coastal habitats, they feed on crustaceans and molluscs. Spur-winged Plovers feed with a slow stalking walk, with their shoulders hunched and head forward.

In New Zealand, Spur-winged Plovers are the most serious avian hazard to aircraft, particularly jet planes, with an unacceptable number of incidents involving strikes or near misses. Birds may congregate on runways and are less agile in flight than gulls. Controls include destruction of nests on airfields, shooting birds on airfields, vehicular bird-scanning patrols on runways before aircraft movement, maintaining unsuitable habitat such as long grass around airfields, and reducing the number of invertebrates on which the birds can feed.

Adult Spur-winged Plover incubating.

Common arctic waders

Eastern Bar-tailed Godwit *Limosa lapponica baueri*
Description Length 37–39 cm; wingspan 62–75 cm.

Bar-tailed Godwits are large waders whose colour, proportions, long legs and long, slightly upturned bill readily distinguish them from other shorebirds like Pied Stilts and Pied Oystercatchers with which they commonly associate in New Zealand. Females are distinctly larger than males, the difference in size being especially noticeable in the length of their bills. In non-breeding plumage their upperparts are mottled brown and grey, the lower back, rump and tail is barred white and brown, and the underparts are dull white, clouded with grey. The legs and feet are black and the bill is black with a pink base. The assumption of breeding plumage begins in January. By the end of February many males have black and buff upperparts, with a brick-red head, neck, breast and underparts. Females are buffy red with fine barring. Adults may still retain varying amounts of breeding plumage when they first arrive in this country at the end of their southward migration. Juvenile birds are like non-breeding adults, but they are more buff in colour with more distinctly mottled upperparts. The barred rump and tail of Bar-tailed Godwits are particularly obvious as birds take flight away from the observer.

Bar-tailed Godwit in breeding plumage.

Population

Bar-tailed Godwits are common arctic migrants. Those that migrate to south-eastern Australia and New Zealand are considered to be from populations which breed in eastern Siberia and Alaska. The East Asian–Australasian Flyway population of Bar-tailed Godwits is an estimated 330,000 birds, of which about 165,000 migrate to Australia. Bar-tailed Godwits are the most numerous arctic wader to visit New Zealand each year, with an estimated 102,000 in this country every summer.

There is an increase in the number of Bar-tailed Godwits throughout New Zealand from the middle of September, with large flocks having returned to favoured localities by the end of that month. National wader counts between 1983 and 1993 show that an annual average of 83,133 Bar-tailed Godwits spent the summer of each of those years in this country. This represents 25 percent of the estimated flyway population. Total numbers counted annually ranged from 66,604 birds in 1993 to 101,771 in 1988, but in most years between 80,000 and 90,000 were recorded during national summer counts. Bar-tailed Godwits are distributed widely around New Zealand during summer, preferring areas with broad intertidal flats, particularly the large northern harbours and Farewell Spit. National wader counts show that three of their favoured localities averaged more than 10,000 Bar-tailed Godwits each summer between 1983 and 1993. Manukau Harbour averaged 15,534, with a highest count of 22,571; Farewell Spit averaged 13,557, with a highest count of 17,181; and Kaipara Harbour averaged 10,381, with a highest count of 14,507. Seven other northern harbours — Firth of Thames, Tauranga, Rangaunu, Ohope/Ohiwa, Parengarenga, Kawhia and Whangarei — each averaged between 3224 and 6479 birds every summer. These nine northern harbours and Farewell Spit between them accommodated about 84 percent of the average number of Bar-tailed Godwits counted in New Zealand in each of the summers of 1983–93. Regular counts show that the total number of Bar-tailed Godwits visiting the Manukau Harbour and Firth of Thames each summer remained more or less the same over the 38-year period from 1960 to 1998.

The main flocks of Bar-tailed Godwits depart New Zealand from about the middle of March until the middle of April. National wader counts between 1984 and 1994 show that an annual average of 12,108 Bar-tailed Godwits spent the winter of each of those years in this country. This is approximately 15 percent of their average summer population during the

Bar-tailed Godwits in non-breeding plumage.

Bar-tailed Godwit preening.

same period. Total numbers counted ranged annually from 6633 in 1987 to 17,172 in 1989. Again, as in summer, the majority were at Manukau and Kaipara harbours and Farewell Spit, which between them averaged 6909 birds or 57 percent of the average winter population.

Behaviour

Bar-tailed Godwits are found on harbours, estuaries and sandy coasts throughout New Zealand, but particularly at inlets and estuaries with broad intertidal mudflats and sandflats. They are very gregarious, feeding in loose groups and roosting in flocks which may number thousands of birds, often in the company of Lesser Knots. They usually choose roosting sites such as sandbanks, shellbanks and spits that are surrounded by water at high tide and therefore afford good visibility in all directions. However, they will also roost on saltmarshes, the edges of shallow coastal lagoons and in coastal paddocks. Like other waders, they have the habit of standing on one leg when resting. They also often sit when resting, but will occasionally stand and stretch a wing and a leg sideways. They fly swiftly, often in compact groups, either with much twisting and turning in unison, or directly in long lines or in V-formation. They are spectacular to

watch when flight after flight is coming in to high-tide roosts. They call frequently during flight. Roosting and feeding groups, particularly if uneasy, often keep up a steady chatter. Most flocks of Bar-tailed Godwits observed departing from Farewell Spit at the commencement of their northward migration left in the evening and on rising tides.

Bar-tailed Godwits usually feed on intertidal mudflats and sandflats, but sometimes on saltmarshes and coastal pastures where the vegetation is short and the ground is wet. Most Bar-tailed Godwits feed along the falling tideline, but they slowly scatter over the flats as the tide recedes further. They obtain food from the surface with rapid stabs of the bill and also from deep in mud and sand by probing. Their food is mainly marine worms, molluscs and crabs.

Bar-tailed Godwits can be aggressive towards other species. A group of Bar-tailed Godwits, assisted by a few Lesser Knots, was seen to tenaciously drive away a small skua which approached their roosting site. Solitary Bar-tailed Godwits are sometimes hotly pursued by an Eastern Curlew. On the other hand, Bar-tailed Godwits tend to resent solitary Eastern Curlews at high-tide roosts and will frequently chase them away by closely flying after them.

Lesser Knot *Calidris canutus rogersi*
Description Length 23–25 cm; wingspan 45–54 cm.

Lesser Knots are medium-sized, robust waders with a short, straight, black bill and short, dull green legs. Adults in non-breeding plumage have plain grey upperparts with paler feather edges, and their underparts are pale grey to off-white. They have some light grey speckling on the neck, breast and flanks and an indistinct white wingbar. Their pale rump is barred white and grey and shows in flight. Lesser Knots begin to moult into their breeding plumage in January. Almost 40 percent of those studied at Farewell Spit showed some breeding plumage by the end of that month. Adults are in full breeding plumage by April. Their head, neck and breast become rusty red, and their back becomes black with rusty and white speckling. Females in breeding plumage are less resplendent than males. Adults may still retain varying amounts of breeding plumage when they first arrive in this country at the end of their southward migration. Juvenile birds are like non-breeding adults, but their back is more scaly with white feather tips and subterminal black lines.

Population

Lesser Knots are common arctic migrants; five subspecies are recognised. They breed in the extreme high arctic and migrate as far south as South Africa, Australia, New Zealand and southern South America. Those that migrate to Australasia are believed to breed on the Chukotski Peninsula in eastern Siberia. The East Asian–Australasian Flyway population of Lesser Knots is an estimated 255,000 birds, of which about 153,000 migrate to Australia. Birds that migrate to New Zealand travel as far as 14,000 to 15,000 kilometres from their breeding grounds. They probably

Lesser Knots in breeding plumage.

cover this considerable distance in a series of three or four long flights. Banding studies have shown that at least some travel to New Zealand by way of Australia. Lesser Knots are the second most numerous arctic wader to visit this country each year, with an estimated 59,000 here every summer.

The first Lesser Knots arrive in New Zealand towards the end of August, with most arriving about the middle of September. National wader counts between 1983 and 1993 show that an annual average of 51,227 Lesser Knots spent the summer of each of those years in this country. This represents 20 percent of the estimated flyway population. Total numbers counted annually ranged from 67,367 birds in 1991 to 33,054 birds in 1993. Although they were widely distributed throughout New Zealand during summer, Lesser Knots actually favoured only a few localities, being several of the northern harbours and Farewell Spit. Two of those localities annually averaged more than 10,000 birds. They were Manukau Harbour and Farewell Spit, which are clearly the summer strongholds of the species in New Zealand. Each summer between 1983 and 1993 they together held an average of about 61 percent of the population of Lesser Knots then in New Zealand. Manukau Harbour averaged 16,083 birds (or 31 percent of the national average), with a highest census count of 22,433, and Farewell Spit averaged 15,538 birds (or 30 percent of the national average), with a highest census count of 24,227. Regular counts show that the number of Lesser Knots recorded in summer at Manukau Harbour increased dramatically over the 38-year period from 1960 to 1998. It averaged 1724 for the period 1960–69; 6540 for the period 1970–79; 15,755 for the period 1980–89; and 18,131 for the period 1990–98. Unfortunately it is not possible, because of the absence of comparable census data, to determine whether this very significant increase is the result of an increase over time in the number of Lesser Knots visiting New Zealand, or is a reduction over time in the number visiting other New Zealand summer locations in preference for Manukau Harbour.

Lesser Knots leave New Zealand from early March until the middle of April, with most departing about the middle of March. National wader counts between 1984 and 1994 show that an annual average of 5310 Lesser Knots spent the winter of each of those years in this country. This is approximately 10 percent of their average annual summer population during the same period. Total numbers counted ranged annually from 2397 in 1984 to 8248 in 1992. By far the majority wintered at Manukau

Lesser Knot in non-breeding plumage.

Harbour, which averaged 3394 birds, or 64 percent of the average winter population.

Behaviour

Lesser Knots are very gregarious. In New Zealand they usually associate with Bar-tailed Godwits, Turnstones, and Pacific Golden Plovers. They often fly in large groups, frequently twisting and turning rapidly. They commonly roost with Bar-tailed Godwits, and usually choose roosting sites such as sandbanks, shellbanks and spits that are surrounded by water at high tide and therefore afford good visibility in all directions. However, they will also roost on saltmarshes, the edges of shallow coastal lagoons, and in coastal paddocks. They can also sometimes be found roosting in large unmixed flocks apart from other waders. Like other waders, they have the habit of standing on one leg when resting. Lesser Knots usually feed at intertidal mudflats and on sandflats, but sometimes on saltmarshes and coastal pastures where the vegetation is short and the ground is wet. They feed on the falling tide with a 'sewing machine' action, rapidly drilling soft mud or wet sand, with their head held low and bill nearly vertical. Their diet while in New Zealand is mainly small molluscs, crustaceans, marine worms, and insects. A study of Lesser Knots at Farewell Spit showed a large increase in the body mass of adults in February before their northward migration in March when they departed with an estimated 'fat' load of about 45 percent. Most flocks of Lesser Knots observed departing from that locality at the beginning of their northward migration left in the evening and on rising tides.

Turnstone *Arenaria interpres*
Description Length 22–24 cm; wingspan 50–57 cm.

Turnstones are stocky, medium-sized waders with a short, slightly uptilted, wedge-shaped bill and short, orange legs. They are distinguished at all times from other medium-sized waders in New Zealand by their stout shape, the contrasting black and white and chestnut colour of their plumage, their short, wedge-shaped, black bill and short, orange legs. Their head and upperparts in non-breeding plumage are dark brown, mottled black and chestnut; their face is variegated black, white and brown; the lower back is white with a white band curving across the upper tail; there is a broad, blackish band on their upper breast, and the rest of their underparts are white. The breeding plumage of males and females differs. Males have black and white patterning over the face and breast, a white cap and nape finely streaked black, rich chestnut and black variegated back and wings, and white abdomen. Females are duller, having a brown wash over much of the head with less distinct black and chestnut patterning. The amount of white on the face and head varies considerably, with some adults in breeding plumage becoming almost white-headed. Newly arrived

Female Turnstone in breeding plumage.

84

juvenile birds are easily distinguished by their dusky heads and duller colouration. Turnstones are very distinctive in flight with their dark upper-parts and striking white wingbars, white lower back and white band on upper tail distinguishing them from all other waders.

Population
Turnstones are common arctic migrants. They breed along the northern coasts and islands of Greenland, Scandinavia, Siberia, Alaska, and the islands of northern Canada. The species is almost cosmopolitan in the northern autumn and winter, being found on the coasts of the Americas, Africa, Madagascar, southern and south-eastern Asia, the islands of the Pacific, and Australasia. The East Asian–Australasian Flyway population

is an estimated 28,000 birds, of which about 14,000 migrate to Australia. Turnstones are the third most numerous arctic wader to visit New Zealand each year, with an estimated 5000 here every summer.

The first Turnstones arrive in New Zealand in the latter half of August, but most arrive in September and October. National wader counts between 1983 and 1993 show that an average of 4227 birds spent the summer of each of those years in this country. This represents 15 percent of the estimated flyway population. Total numbers counted ranged annually from 5915 in 1984 to 2394 birds in 1990. Turnstones occur throughout New Zealand, more or less evenly distributed between both main islands although rarely on the western coast of the South Island. They concentrate in flocks of various sizes in certain favoured coastal localities which tend to be the northern harbours, Nelson-Marlborough region, and southern estuaries. National wader counts show that in the summers of 1983–93, Parengarenga Harbour and Farewell Spit together held an average of about 42 percent of the Turnstones then in New Zealand. Parengarenga Harbour averaged 915 birds (or 22 percent of the national average), with a highest census count of 1500, and Farewell Spit averaged 846 birds (or 20 percent of the national average), with a highest census count of 1792. There were also several hundred on average every summer at a number of other localities, particularly Kaipara and Manukau harbours, and Invercargill Estuary. However, small numbers of Turnstones are likely to turn up at any coastal

Turnstones in breeding plumage with Pied Oystercatchers behind.

lagoon or estuary, especially when they first arrive in this country at the end of their southward migration.

Most Turnstones depart New Zealand in March and April, but some do not leave until May. National wader counts between 1984 and 1994 show that an annual average of 586 Turnstones spent the winter of each of those years in this country. This is approximately 14 percent of their average summer population during the same period. Total numbers counted ranged widely from 83 in 1990 to 1436 in 1992. The majority wintered at Farewell Spit, which averaged 176 birds (or 30 percent of the average winter population), and at Parengarenga Harbour which averaged 165 birds (or 28 percent of the average winter population). These wintering birds are probably mainly young individuals from the previous breeding season, the highly variable numbers perhaps reflecting breeding success in the immediately preceding northern summer.

Behaviour

Turnstones do not move inland to any great distance, but they sometimes visit rough farmland near the sea. They tend to avoid smooth, sandy beaches and open mudflats and sandflats, preferring to feed among rockpools or on exposed rocky reefs, on shelly or stony foreshores, and on saltmarsh flats. They busily fossick along the tideline of foreshores, picking at stranded debris, probing under it, pushing it aside, or flicking it and stones over in search of food items such as sandhoppers. They will dash quickly between waves to peck at various items and to search through debris. They prey on small mud crabs, which they take by both chasing them on the surface and by digging them from their burrows. Turnstones are among the last birds to arrive at a high-tide roost. Groups often roost densely together, near to but apart from other resting birds. Lone individuals will readily join groups of other small waders. Pacific Golden Plovers and Sharp-tailed Sandpipers will often roost on the edge of a Turnstone flock, and occasionally Lesser Knots are found with them. A marked feeding and resting association of Turnstones and Pacific Golden Plovers has often been noticed. Turnstones sometimes begin to feed as soon as the tide starts to ebb while most other waders continue to roost. They are capable of being particularly aggressive. A group of Turnstones has been observed chasing and attacking a badly emaciated individual of their own kind, and would almost certainly have killed it had it not been for human intervention.

Pacific Golden Plover *Pluvialis fulva*
Decription Length 23–26 cm; wingspan 60–72 cm.

Pacific Golden Plovers are medium-sized plovers with large, black eyes, a short, black bill and long, generally dark grey legs in which the sexes are alike, but the breeding and non-breeding plumages are very different. In non-breeding plumage they are mottled brown and buff above with a golden suffusion which becomes more obvious as breeding plumage is assumed; their throat and breast are grey and yellowish-buff; the eyebrow is pale buff, and the underparts buffy white. Some adults begin to show black on the underparts in February, and the moult into breeding plumage is complete by April when they leave New Zealand. Adults are striking in full breeding plumage. They have a white forehead and brownish upperparts heavily speckled golden yellow and white. Their black face and underparts are separated from the upperparts by a broad white stripe running above the eye and down the side of the body to the flanks. Adults may still retain varying amounts of breeding plumage when they first arrive in this country at the end of their southward migration. Juvenile birds are like non-breeding adults but have more heavily speckled upperparts and a more yellow head and underparts, with more dark markings on the breast and flanks. In flight, Pacific Golden Plovers have a dark brown upperwing with an indistinct pale wingbar and uniform brownish-grey underwing and armpits.

Population
Pacific Golden Plovers are common arctic migrants. They breed on the arctic and subarctic tundra of Siberia and western Alaska. They migrate south on a broad front to many countries including Australasia and most Pacific islands. The East Asian–Australasian Flyway population is an estimated 90,000 birds, of which about 9000 migrate to Australia. Pacific Golden Plovers are the fourth most numerous arctic wader to visit New Zealand each year, with an estimated 650 here every summer.

The earliest Pacific Golden Plovers arrive in New Zealand in September, but most arrive from October to early November. National wader counts between 1983 and 1993 show that an annual average of 466 Pacific Golden Plovers spent the summer of each of those years in this country. This represents only 0.5 percent of the estimated flyway population. The number actually present in New Zealand can vary considerably

Pacific Golden Plovers and Terek Sandpiper.

from year to year even within the same locality. Total numbers counted annually ranged widely from only 30 birds in 1984 to 1120 in 1986. Substantial differences in the summer numbers of Pacific Golden Plovers visiting Manukau Harbour and the Firth of Thames combined have also been noted between decades, especially between the 1970s and 1990s. Overall numbers dropped from an annual summer average of 113 birds between 1970 and 1979 to an annual summer average of only 33 birds between 1990 and

Pacific Golden Plover moulting.

1998. In the four-year period between 1995 and 1998, the annual summer total did not exceed 25 birds over the two harbours combined. However, because of their irregular habits as described in the next section, Pacific Golden Plovers are not one of the best subjects for census work, and many may sometimes have been overlooked.

Pacific Golden Plovers are widespread during summer at a number of harbours and estuaries and some lakes throughout New Zealand. National wader counts between 1983 and 1993 show that Parengarenga Harbour was the most favoured summer locality in New Zealand, with an annual average of 75 birds (or 16 percent of the national average), and a highest census count of 200. Other favoured localities were Kaipara and Manukau harbours and Lake Wairarapa in the North Island, and Lake Ellesmere and Invercargill Estuary in the South Island. Other suitable places, such as Manawatu Estuary, also have good numbers each year. Some Pacific Golden Plovers even regularly spend all of their time in New Zealand at seemingly unsuitable localities, such as the rocky coastline at the mouth of the Waiongana Stream near New Plymouth where between 5 and 13 birds have been resident over recent summers. It is entirely possible that many Pacific Golden Plovers spend their New Zealand summers unnoticed in similar localities throughout the country.

Pacific Golden Plovers very seldom spend the winter in New Zealand. Virtually all seem to depart this country in late March and early April. National wader counts between 1984 and 1994 show that an annual average of only 3 spent the winter of each of those years here, with the highest count being 7 in 1994.

Behaviour

When Pacific Golden Plovers first arrive in New Zealand they are often found scattered singly or in very small groups, whereas later they gather together in larger flocks. They are irregular in their habits. A flock may be found on a beach or a mudflat one day and the next will be some distance inland on short pasture or ploughed ground where it can easily be overlooked. They do not form compact groups when roosting but are usually somewhat scattered. They can be very mobile over the high-tide period and have no regular roost. They will often not use the same high-tide roosts as other waders. However, a marked feeding and resting association of Pacific Golden Plovers and Turnstones has often been noticed. In New Zealand, they are normally alert and shy and difficult to approach closely. When alert they stand tall, looking slim with their long neck and large head. They are usually silent on the ground, but give a clear and melodious call in flight. They fly swiftly and directly in groups, with strong and regular wingbeats. Pacific Golden Plovers eat a wide range of insects and their larvae, spiders, earthworms and plant seeds, as well as small crustaceans such as crabs, molluscs and marine worms.

Pacific Golden Plover in breeding plumage.

91

Red-necked Stint *Calidris ruficollis*
Description Length 13–16 cm; wingspan 29–33 cm.

Red-necked Stints are the smallest of the arctic waders which regularly visit New Zealand. They are dwarfed even by the Wrybill Plovers and Banded Dotterels with which they commonly associate. In non-breeding plumage Red-necked Stints are predominantly grey and white. The crown, sides of the face, hindneck and upperparts are pale grey with a brownish tinge. The larger feathers have pale edges with black shafts. They have an indistinct white forehead and superciliary stripe. The chin, throat and

belly are white. The breast has variable amounts of grey flecking. They have a short, stubby, black bill and short, black legs. In flight, there is a distinct white bar on the upperwing, and white sides to the rump and upper tail contrast with a black stripe down their centre. Adults assume breeding plumage before commencing northward migration. The sides of the face, chin, throat and neck become brick-red. The rest of the under-parts remain white, except for some dusky flecking on the sides of the breast. The feathers of the crown and back become blackish-brown with rufous margins. The wing coverts are also blackish-brown, the greater coverts being tipped with white. Adults may still retain varying amounts of breeding plumage when they first arrive in this country at the end of their southward migration. Juvenile birds are like non-breeding adults, but their crown and sides of the breast are washed pale rufous, and their back feathers are dark rufous-edged, which contrasts with their greyer pale-edged wing coverts.

Population

Red-necked Stints are common arctic migrants. They breed in northern Siberia and in north-western Alaska and migrate to Malaysia, the Philippines and Australasia. They are the most abundant migrant wader in Australasia. The East Asian–Australasian Flyway population is an esti-mated 471,000 birds, of which about 353,000 migrate to Australia, partic-ularly south-eastern Australia. Red-necked Stints are the fifth most numerous arctic wader to visit New Zealand each year, but in numbers vastly fewer than go to Australia. It is estimated that about 175 are in this country every summer.

The first Red-necked Stints arrive in New Zealand in late September, with most reaching this country during October and November. National wader counts between 1983 and 1993 show that an average of only 158 birds, with a maximum of 231 in 1991, were in New Zealand each summer during that period. Red-necked Stints were consistently most numerous at Lake Ellesmere. Up to 125 at one time, and 43 percent of the New Zealand summer average, were recorded there. Lake Ellesmere is clearly their most favoured locality in New Zealand. Awarua Bay, Manukau Harbour and Farewell Spit are also important localities, and a few are reported annually from other suitable places such as Manawatu Estuary.

Red-necked Stints depart on northward migration from late March. National wader counts show that an average of only 26, with a maximum of 56 in 1988, stayed in New Zealand during each of the winters of 1984–94.

Behaviour

In New Zealand, Red-necked Stints usually associate with Banded Dotterels and Wrybill Plovers in localities where those species are present. Red-necked Stints can frequently be located by examining flocks of wintering Wrybill Plovers at their high-tide roosts. Red-necked Stints, Curlew Sandpipers, and Sharp-tailed Sandpipers also often associate. It may be no coincidence that in this country the largest numbers of both Red-necked Stints and Curlew Sandpipers are regularly to be found at Lake Ellesmere. Red-necked Stints are voracious feeders, mainly at tidal mudflats, sandflats, saltmarshes and on the margins of coastal lagoons, where they busily run about probing and pecking with a rapid 'sewing-machine' action. Virtually nothing is known about their diet in New Zealand, but in Australia they mostly feed on small marine invertebrates, molluscs, crustaceans and insects. They are swift and agile fliers and tend to flutter on landing.

Whimbrel *Numenius phaeopus*
Description Length 40–45 cm; wingspan 76–89 cm.

Whimbrels are about the same size as Bar-tailed Godwits but look darker and more robust and have a long (9 cm), distinctly downcurved, black bill. Whimbrels are almost completely steaked greyish-brown and buff. Their crown is dark brown with a clear pale stripe down the centre, which is very distinctive when seen from the front. The Eastern Curlew does not have a stripe like this down the centre of its crown. Whimbrels have bluish-grey legs.

The two subspecies of Whimbrel which occur in New Zealand, the Asiatic Whimbrel and the American Whimbrel, are normally indistinguishable on the ground. However, the Asiatic Whimbrel has a lower back and rump which are whitish barred with brown in contrast to the uniform brown of the rest of its upperparts, whereas the American Whimbrel has a brown lower back and rump which is uniform in colouration with the rest of its upperparts. These differences in the colouration of the lower back and rump are the only features by which the two subspecies can be separated in the field. It is not safe to identify a Whimbrel to subspecies unless they are clearly seen. Their differences should be looked for when a Whimbrel is in flight, the whitish lower back and rump of the Asiatic Whimbrel showing as a pale triangle pointed up the lower back. However, it is not always possible to see the relevant features of a Whimbrel in flight, and accordingly, many Whimbrels cannot be identified to subspecies.

Population

Whimbrels are common arctic migrants. The Asiatic Whimbrel, which breeds in eastern Siberia, is the subspecies more commonly found in Australasia. The East Asian–Australasian Flyway population of this subspecies is an estimated 40,000 birds, of which about 10,000 migrate to Australia. The American Whimbrel breeds in northern North America and visits New Zealand regularly in small numbers. Only 42 percent of the Whimbrels recorded at Manukau Harbour and the Firth of Thames between 1960 and 1998 were identified to subspecies. Of these, 90 percent were Asiatic Whimbrels and 10 percent were American Whimbrels. The following figures are of both Whimbrel subspecies combined. Whimbrels are the sixth most numerous arctic wader to visit New Zealand each year. It is estimated that about 120 are in this country every summer.

Single Whimbrels or small flocks make their New Zealand landfall almost anywhere along the coast, but soon move to tidal estuaries and harbours. National wader counts show that an annual average of 89 spent the summers of 1983–93 in this country. The number counted during censuses in those years ranged widely, from 33 in 1990 to 178 in 1992. The largest flocks reported were of 46 and 53 birds at Parengarenga Harbour in 1986 and 1992 respectively. During national wader counts, most Whimbrels were reported from Parengarenga Harbour (the most favoured site), the Firth of Thames, Kaipara Harbour and Farewell Spit. Those four localities together accounted for an annual average of 76 birds, or 85 percent of the average number of Whimbrels which spent each of the summers of 1983–93 in New Zealand.

National wader counts show that a high proportion of Whimbrels (an annual average of 24, being about 27 percent of the summer average) stayed in New Zealand during the winters of 1984–94. Again, numbers ranged widely from 70 in 1985 to 7 in 1993.

Behaviour

Whimbrels are gregarious, and individuals of both subspecies sometimes associate temporarily. A single Whimbrel will often be found in the company of Eastern Curlews when that species is present. Whimbrels usually feed and roost in small groups. They are often very wary, standing alert at

the edge of other roosting waders such as Bar-tailed Godwits and Lesser Knots, and will take flight long before a human can get anywhere near them. A single roosting bird can be almost impossible to find among Bar-tailed Godwits until it lifts its head and shows its down-curved bill. Sometimes Whimbrels call repeatedly with an even rippling whistle of about seven notes. This call can be the first indication there is a Whimbrel among a passing flock of Bar-tailed Godwits. Whimbrels feed mainly on crabs, which they dismember, and marine worms which they pick from the surface or take by probing.

Curlew Sandpiper *Calidris ferruginea*
Description Length 18–23 cm; wingspan 38–41 cm.

Curlew Sandpipers are small, slim sandpipers with a long, thin, down-curved, black bill, longish black legs, white rump and a narrow, white wingbar. The white rump and wingbar are evident in flight. Non-breeding Curlew Sandpipers have a pale grey-brown head and upperparts, white superciliary stripes, and white underparts with a pale grey wash and dark streaks on the breast. The time of assumption of breeding plumage varies considerably. Some adult Curlew Sandpipers begin to redden in January, and by the end of March are in full breeding plumage. However, there is considerable variation in the plumage colour of Curlew Sandpipers in autumn. Some are well coloured, while others have only some colour, or are pale. Curlew Sandpipers in full breeding plumage are among the most colourful of waders. Males are particularly brilliant with rich rufous on the sides of the head and on the breast. The upperparts are dark to reddish brown with whitish edges to dark-centred feathers, giving a spangled effect. The crown has a rich waved pattern running laterally. The whole effect is enhanced by touches of white on the sides, on the edges of the dark brown wing quills and on the upper tail coverts, which are barred with black. Adults may still retain varying amounts of breeding plumage when they first arrive in this country at the end of their southward migration. Juvenile birds are like non-breeding adults, but their back feathers

Curlew Sandpipers feeding with Lesser Knot.

are darker with white edges, and their lightly streaked neck and breast is washed buff.

Population

Curlew Sandpipers are common arctic migrants. They breed in central Siberia, although a few have bred in Alaska, well to the east of their normal range. They spend the northern winter in Africa, southern Asia and Australasia. The East Asian–Australasian Flyway population is an estimated 250,000 birds, of which about 188,000 migrate to Australia. Curlew Sandpipers are the seventh most numerous arctic wader to visit New Zealand each year, but in numbers vastly fewer than go to Australia. It is estimated that about 85 are in this country every summer.

Curlew Sandpipers are among the first arctic waders to arrive in New Zealand at the end of the northern summer. Some newly arrived birds have been seen in the Firth of Thames as early as 5 September. National wader counts between 1983 and 1993 show that an average of only 75 were present in New Zealand each summer during that period, with a maximum of 136 being counted in the summer census of 1992. Large flocks of Curlew Sandpipers are exceptional, but a considerable number can sometimes be found in one locality. The 90 at Parengarenga Harbour in the summer of 1991–92 is probably the largest number ever recorded from one locality in New Zealand. Forty-seven were seen together on the Karaka shellbanks in January 1995, and 70 were counted there in October of the same year. Small numbers of Curlew Sandpipers are regularly reported from many estuaries, but the species is consistently present each year in greatest numbers at Lake Ellesmere. In national wader counts between 1983 and 1993, Lake Ellesmere recorded 40 percent of the New Zealand summer average of Curlew Sandpipers, with a maximum of 59 being counted there at one time. The Firth of Thames, Parengarenga Harbour and Awarua Bay are also favoured localities.

Curlew Sandpipers are frequently very late leaving New Zealand on northward migration at the end of the southern summer. Individuals in full breeding plumage can still be seen busily feeding on tidal flats until as late as mid-May, long after other arctic migrants have left for their breeding grounds. Nevertheless, very few Curlew Sandpipers normally remain in New Zealand during the southern winter. An annual average of

Right: Curlew Sandpiper in breeding plumage.

12 were reported in this country during national wader counts in the winters of 1984–94. However, this average is about double what it would have been if the quite exceptional number of 88 Curlew Sandpipers had not been recorded during the winter census of 1992. The next highest winter census counts were of only 11 birds in both 1988 and 1989.

Behaviour

Curlew Sandpipers favour tidal flats, brackish pools and the margins of some coastal lakes and lagoons. They tend to associate mainly with Wrybills where they occur in the northern harbours, and with Banded Dotterels on the shores of Lake Ellesmere. One or two Curlew Sandpipers in non-breeding plumage can be very difficult to detect when they are roosting among large groups of densely packed Lesser Knots as they sometimes do at places like Manawatu Estuary. They feed in bare wet mud or in shallow water, both probing deeply with their long bills and picking from the surface. Little is known of their diet in New Zealand, but they appear to feed mainly on marine worms, molluscs and crustaceans. Groups of Curlew Sandpipers are particularly spectacular in flight. They fly very swiftly with high twittering calls and twist and turn sharply in all directions.

Curlew Sandpiper in non-breeding plumage.

Sharp-tailed Sandpiper *Calidris acuminata*
Description Length 17–22 cm; wingspan 36–43 cm.

Sharp-tailed Sandpipers are brownish, medium-sized waders that are very similar in size, shape, general appearance and plumage to Pectoral Sandpipers. In the field, they are separable from that species only after close and careful scrutiny. The brown colouration of Sharp-tailed Sandpipers makes them conspicuous when roosting among greyish Wrybill Plovers as they often do at Manukau Harbour and the Firth of Thames. Individuals can vary considerably in appearance depending on age and season. The sexes are alike. Adults in non-breeding plumage have a rufous crown that is streaked black, with whitish superciliary stripes that vary in clarity. Their upperparts are dark brown with pale feather edges, and their rump and tail are dark brown with white sides. Their neck and breast are mottled grey or buffish with irregular dark streaks, but in full non-breeding plumage they are often without streaking. The breast fades to white on the abdomen, but there is no sharp line of demarcation between the breast and the abdomen as there is in the Pectoral Sandpiper. This is the principal character by which Sharp-tailed

Sandpipers in non-breeding plumage can be distinguished from Pectoral Sandpipers in non-breeding plumage in the field. Their dark bill is slightly less decurved than that of the Pectoral Sandpiper. This difference is noticeable in the field. The legs are yellowish-green. The upperparts of Sharp-tailed Sandpipers in breeding plumage are of a rich chestnut colouration with buff feather edges. This is very noticeable in the field. Their neck and breast become heavily streaked, and small, dark, crescentic streaks develop on the lower breast and flanks. These crescentic streaks are absent from Pectoral Sandpipers in breeding plumage. Juvenile Sharp-tailed Sandpipers have broad, pale superciliary stripes bordering a very distinct, bright rufous cap. Their foreneck and upper breast are orange-buff, with a narrow gorget of fine streaks across the upper neck. In flight, Sharp-tailed Sandpipers have a faint white wingbar and white sides to their dark rump and upper tail.

Population

Sharp-tailed Sandpipers are common arctic migrants. They breed in north-eastern Siberia. The East Asian–Australasian Flyway population of the species is an estimated 166,000 birds, almost all of which migrate to Australia. Sharp-tailed Sandpipers are the eighth most numerous arctic

Sharp-tailed Sandpipers in non-breeding plumage.

wader to visit New Zealand each year, but in numbers vastly less than go to Australia. It is estimated that about 80 are in this country every summer.

National wader counts between 1983 and 1993 show that an annual average of only 68 spent the summers of those years in this country, the most being 173 recorded in 1987. They were reported from estuaries or lakes throughout the country, with average summer numbers being highest at the Firth of Thames (13), followed by Lake Ellesmere (11) and Invercargill Estuary (11). The largest number of birds reported from any one locality during national wader counts was 48 at Lake Ellesmere. Some other suitable localities, such as Manawatu Estuary, are favoured with small numbers of resident Sharp-tailed Sandpipers each summer. It appears that very few spend the winter in New Zealand. None were reported from seven of the national wader counts held in the winters of 1984–94, and the species was recorded on very few occasions over many years during a large number of separate winter counts at Manukau Harbour and the Firth of Thames.

Behaviour

Sharp-tailed Sandpipers are often found in small flocks. By New Zealand standards a 'big' flock of Sharp-tailed Sandpipers would contain more than 30 birds. They tend to keep to themselves, but on tidal mudflats may associate with Lesser Knots, Curlew Sandpipers, or Red-necked Stints. Sharp-tailed Sandpipers sometimes roost among other small waders such as Wrybills on shellbanks or sandbanks, but a group of them will frequently be found resting apart from other birds. They will feed on tidal mudflats well back from the tideline, but they seem to prefer coastal lakes, lagoons and pools where they often feed and roost around the edges, almost if not quite hidden, among glasswort, batchelor's button, and other similar vegetation. In this situation, such as on the Stilt Pools at Miranda, they can be hard to find and therefore may easily escape

Sharp-tailed Sandpiper in breeding plumage.

notice. Sharp-tailed Sandpipers are omnivorous, and will eat seeds, worms, molluscs, crustaceans and insects. They are usually silent in New Zealand, but may call when disturbed into flight.

Eastern Curlew *Numenius madagascariensis*
Description Length 60–66 cm; wingspan about 110 cm.

Eastern Curlews are the largest of the arctic waders that visit New Zealand each summer. They are a spectacular bird with a distinctive, very long (19 cm), downcurved bill. Their whole body is streaked greyish-brown and buff which is paler on the underparts, but they do not have a clear pale stripe down the centre of the crown as Whimbrels do. They have an indistinct pale superciliary stripe and a brown rump. The bill is dark brown with a pink base to the lower mandible, and their legs are bluish-grey. The combination of large size, long downcurved bill, and uniformly patterned plumage readily distinguish Eastern Curlews from all other waders in New Zealand.

Population

Eastern Curlews are common arctic migrants. They breed in north-eastern Asia. The East Asian–Australasian Flyway population of the species is an estimated 21,000 birds, of which about 19,000 migrate to Australia. Eastern Curlews are the ninth most numerous arctic wader to visit New Zealand each year, but in numbers very much fewer than go to Australia. It is estimated that about 35 are in this country every summer.

Eastern Curlews may stop on almost any stretch of coast, usually the west, when they first arrive in New Zealand in about late September, but they soon move to localities most frequented by other waders. Although they are widespread in New Zealand during summer, most Eastern Curlews are regularly reported from just three sites: Manukau Harbour, the Firth of Thames and Farewell Spit. An annual average of only 29 birds were recorded in national wader counts during the summers of 1983–93. The largest number so far recorded here at one time was 37 birds seen at Farewell Spit in September 1962. Farewell Spit has always been a locality favoured by Eastern Curlews while in New Zealand. For instance, 35 were seen there in January 1967, and a flock of 15 was recorded in February 1978. However, the number annually present at Farewell Spit has clearly been declining in recent years. Only between 5 and 13 birds (an average

Eastern Curlew in flight showing moulting wing feathers.

of 8 per annum) were recorded there in national wader counts during the summers of 1983–93. Those counts show that, during the same period, the number of Eastern Curlews annually recorded in New Zealand generally has also been declining. From 1983 to 1986, between 26 and 46 (an average of 38 birds each year) were counted. Between 1987 and 1990 that number dropped to 24–29 (an average of 27 birds each year), and between 1991 and 1993 it dropped even further to 19–22 (an average of 21 birds each year). Despite this overall trend, the number of Eastern Curlews annually visiting Manukau Harbour and the Firth of Thames combined did not change between 1960 and 1998, but in recent years more have annually visited Manukau Harbour while fewer have visited the Firth of Thames.

Most Eastern Curlews leave New Zealand in about late March, but national wader counts showed that a high proportion (an annual average of 7, being about 24 percent of the summer average) stayed in this country during the winters of 1984–94. Numbers recorded ranged between 1 in 1991 and 22 in 1985.

Behaviour

Eastern Curlews are normally birds of the tideline, feeding mainly in tidal estuaries and harbours, but they will also feed in muddy coastal lagoons. They may also feed and rest among and sometimes actually under low-growing mangroves. Their diet is mostly crustaceans, particularly ghost shrimps and mud crabs, and marine worms, which they obtain by probing deeply into mud or by picking at the surface. Eastern Curlews associate with Bar-tailed Godwits, Lesser Knots and Pied Oystercatchers at high-tide roosts, but tend to keep slightly apart from other waders and so are easily noticed. They are frequently the last waders to arrive at a high-tide roost. Instead of proceeding directly to dry land, they will often stand in shallow water until eventually the rising tide forces them out. Sometimes they will resume feeding as the tide drops, at other times they will continue resting long after high tide. They are very wary and difficult to approach when both feeding and roosting, and are easily put to flight.

Eastern Curlews are normally silent in New Zealand, but they have a distinctive call given mostly in flight which is audible at a considerable distance. Eastern Curlews are sometimes aggressive towards other birds feeding near them. They have been seen to run at and chase away Bar-tailed Godwits and Red-billed Gulls. An Eastern Curlew will sometimes hotly pursue a solitary Bar-tailed Godwit in flight. On the other hand, Bar-tailed Godwits tend to resent solitary Eastern Curlews at high-tide roosts and will sometimes chase them away by closely flying after them. An Eastern Curlew will call loudly when being pursued in this manner, and this may be the first indication a human has of its presence.

Common gulls and terns

Three species of gull are resident and breed in New Zealand. Black-backed Gulls are the largest. They and Red-billed Gulls are common, widely distributed, and familiar birds found mostly in coastal areas throughout New Zealand, frequently in association with humans and their activities. Black-billed Gulls are less common and less widely distributed being very infrequently or never found in some coastal areas of New Zealand. They breed mainly on the inland riverbeds of the eastern South Island. They are found less frequently than the other gulls in association with humans and their activities. Black-backed and Red-billed Gulls are more aggressive, predatory and scavenging by nature than are Black-billed Gulls. All three gulls are conspicuous and often numerous inhabitants of many shorebird localities in New Zealand.

Four species of tern are resident and breed on the New Zealand mainland. Three are common, and one, the Fairy Tern, is a critically endangered species in this country. White-fronted Terns are by far the most common and are characteristic birds of the New Zealand coasts. Caspian Terns are the largest and are also mainly a coastal species. Black-fronted Terns are found on some coasts during part of the year, but breed on the inland riverbeds of the eastern South Island. All three common species can be regularly met with at many shorebird localities in New Zealand.

Southern Black-backed Gull *Larus dominicanus dominicanus*
Description Length 49–62 cm; wingspan 106–142 cm.

Southern Black-backed Gulls (commonly known as Black-backed Gulls) are the largest gulls in New Zealand. The subspecies present in this country breeds widely in the subantarctic and temperate southern hemisphere. Black-backed Gulls pass through a complicated succession of plumage changes in their progression from juvenile to adult. The plumage changes progressively during their second and third years until the black and white adult plumage is attained, usually in the fourth year. Juveniles are dull brown, with pale feather edges over much of the body giving them a mottled appearance. The bill and eyes are dark brown and legs pinkish-brown. In the second year, the head, neck and underparts are white, with brown mottling and flecking. The back and upperwings are brown and black. The bill is a dull yellowish or greenish colour and darker at the tip. The legs are pinkish-brown to greyish-green. In the third year,

Adult Black-backed Gulls.

the head and underparts are white and the neck is lightly flecked with brown. The back and upperwings are brown and black. The rump and tail are white, and the tail has a black band across its tip. The bill is dull yellow, darker at the tip, and the legs are yellowish-green. In adult plumage they have a white head, neck, underbody, rump and tail. The underwings are white except for the tips, which are black. The back is black, as are the upperwings, which have a narrow, white trailing edge. The bill is heavy and yellow with a red spot near the tip of the lower mandible. The gape and eyelid are orange, the eye is pale yellow, and the legs are greenish-yellow. The sexes are alike in plumage. Males are usually slightly larger than females, but this difference is not noticeable in the field.

Population

Black-backed Gulls are a native species, but they are not protected. They are the most abundant and widespread gull in New Zealand. They live in a wide variety of habitats from coastal to alpine areas. There must be very few stretches of the New Zealand coastline without at least one resident pair. The total population of Black-backed Gulls in the New Zealand region was estimated to be over one million breeding pairs in the early

1980s. They are an aggressive and successful species which no doubt benefited significantly from the additional food supplies that became readily available to them with the advent of human facilities like whaling stations, meat and fish processing plants, fishing boats, rubbish dumps and sewer outfalls. However, there are no longer any whaling stations, offal and sewage discharges into the sea have been greatly reduced, and improved waste management has reduced the amount of refuse available from rubbish dumps. Because of this, the number of Black-backed Gulls may have declined in some areas, but the total New Zealand population must still be very significant. The small number periodically shot at some rubbish dumps, and the nesting attempts destroyed and birds shot in the interest of aircraft safety, are not likely to have any effect on the overall population of the species.

Juvenile Black-backed Gull.

Behaviour

Black-backed Gulls are generally gregarious. They usually nest in colonies, and frequently feed and roost in large groups. Black-backed Gulls leave their nesting colonies during summer and autumn when there is a rapid growth of flocks at feeding sites and nearby roosts. They mix freely with the smaller gulls and terns on the coast, and can be seen congregating with Red-billed Gulls at rubbish dumps. Black-backed Gulls are

mainly sedentary, but groups will commute considerable distances each day between roosting sites near the coast and rubbish dumps and inland feeding areas. They rarely venture very far out to sea. They take an exceptionally wide range of food such as offal and refuse at rubbish dumps, sewer outfalls and behind boats, carrion on roads, the fresh or putrefying carcases of whales, seals, dead fish and birds cast up on the beach, marine invertebrates, small fish and shellfish along the coastline and in coastal waters, worms and insects in wet or freshly ploughed pasture in farmlands and wet playing fields and parks in coastal towns and cities, placentas and dead lambs in lambing paddocks, and they have even been known to eat soft parts of disabled old sheep and of very young or weak lambs. They can often be seen on the coast repeatedly carrying a mollusc aloft and dropping it on hard sand or on rock in order to break it open. Black-backed Gulls sometimes 'paddle' with their feet, usually when they are standing in very shallow water or on mudflats or exposed beds of seaweed, but sometimes on land, presumably in order to make invertebrate food items reveal themselves. They will take beetles, frogs, lizards, mammals and eat small fruit and other plant material. They have been known to take tuatara and young Fairy Prions on Stephens Island. Black-backed Gulls predate the eggs of other species, including gannets, and take the chicks of Northern New Zealand Dotterels, Red-billed and Black-billed Gulls, and White-fronted, Black-fronted and Caspian Terns near which they frequently nest. For this purpose, Black-backed Gulls often 'patrol' nesting shorebird colonies. During 'patrolling', they fly with their beak pointed towards the ground and scan the area with sharp sweeping movements of the head. They will even predate the young of their own species, and an adult has been seen near a coastal nesting colony repeatedly carrying a young one aloft and dropping it onto hard sand as they do with molluscs. Black-backed Gulls can sometimes be seen hesitating in flight while they scratch their head with a foot.

Black-backed Gulls nest between October and February, usually in colonies which may sometimes number several thousand pairs, in a wide range of habitats such as on coastal dunes, sandspits, boulderbanks, gravel beaches, islands and rocky islets, lakeshores and riverbeds. They have been known to nest in colonies some distance inland at a considerable altitude in the mountains. Some nest as solitary pairs or small groups on beaches, coastal rock stacks and cliffs, on logs stranded in river estuaries, near high mountain tarns, and quite frequently on the roofs of city buildings in some areas. Nesting Black-backed Gulls usually vigorously defend their nest site or nearby young from human intruders by repeated aerial attacks accompanied by much loud calling. Nesting gulls are most aggressive towards other Black-backed Gulls venturing near their nests just before eggs hatch and while chicks are small. A number, mainly males, die from fatal wounds inflicted in fights at this time.

Red-billed Gull *Larus novaehollandiae scopulinus*
Description Length 36–44 cm; wingspan 91–96 cm.

Red-billed Gulls are small, slender, grey and white gulls. They are very similar in general appearance to Black-billed Gulls, but the adult of that species has a longer and more slender black bill, normally black legs and

Adult Red-billed Gull in breeding plumage.

feet, and a different upper wingtip pattern. Adult Red-billed Gulls are readily identified by their stoutish, bright red bill, bright red legs and feet, and their distinctive dark-tipped upper wings in flight. The sexes are alike in plumage. Males are usually slightly larger than females, but this difference is not noticeable in the field. The entire plumage of the head, neck, underparts and tail is white. The mantle, back and wing coverts are uniformly pale grey, as are the underwing coverts. They have distinctive dark-tipped upper wings in flight because the visible parts of the first and second primaries are mainly black, with a white patch or 'mirror' near the tip, and the terminal portion of the third to sixth primaries is black. All are only very narrowly tipped white. They have a bright red eye-ring and white iris. Their gape and mouth are red. The red colouration of bare parts is duller outside the breeding season. Separation of juvenile and immature Red-billed Gulls from juvenile and immature Black-billed Gulls in the field in areas where both species can occur is notoriously difficult, particularly when they are at rest. Young birds of both species can look almost identical. Initially, the upper wing coverts of both are heavily spotted brownish. Those spots gradually disappear, but young Black-billed Gulls lose them sooner than young Red-billed Gulls. The amount of black in the upper wingtips of young Black-billed Gulls can vary with age. Birds of both species go through a poorly known process of change with age in the colour of their bare parts, particularly in the colouration of the bill and legs. This confusing process may be further complicated by seasonal and individual variation. There are times when the colouration of the bill and legs can be almost identical in the young of both species. Despite these difficulties, Black-billed Gulls always have a longer and thinner bill than Red-billed Gulls. The shape and length of the bill, if it can be seen well enough, may be the only reliable character by which the young of these two species can be safely distinguished in the field.

Population

Red-billed Gulls are a protected native species. They are very common along the coasts of New Zealand and rarely venture more than a few kilometres inland. They are, however, more common on the eastern coasts of the North and South islands where they spend much time foraging over inshore waters. Nesting colonies are also found mainly on those coasts. The total population of Red-billed Gulls in New Zealand was estimated to be over 100,000 breeding pairs in the early 1980s. Like the Black-backed

119

Gull, it is a successful species that no doubt benefited significantly from the additional food supplies which became readily available with the advent of human facilities like rubbish dumps and sewer outfalls. However, sewage discharges into the sea have been greatly reduced, and improved waste management has reduced the amount of refuse available from rubbish dumps. Because of this, the number of Red-billed Gulls may have declined in some areas. Black-backed Gulls will take the eggs and chicks of Red-billed Gulls. Stoats take eggs from their nests and eat the contents. Ship rats prevented them nesting on an offshore islet near New Plymouth. Nevertheless, despite these and other problems, Red-billed Gulls remain plentiful and there is no reason to believe that their overall population has declined.

Behaviour

Red-billed Gulls are predominantly a coastal species. They breed throughout New Zealand, almost exclusively on the coastline or offshore islands, between October and February in densely packed colonies of varying sizes. They nest mostly on sandspits, boulderbanks, shellbanks, gravel beaches, rocky headlands and rocky islets, usually near areas of marine upwelling rich in plankton. Inland breeding is quite exceptional, but there is a long-established colony at Lake Rotorua where they nest with Black-billed Gulls. The number of birds at colonies may start building up at any time from late July onwards, but some colonies are not fully occupied until October. They begin to disperse from breeding colonies in mid-January. Many undertake regular seasonal movements between their breeding colonies and traditional wintering areas. For example, birds from the large nesting colony at Kaikoura winter in Wellington Harbour, and have been found as far away as Auckland and Invercargill.

Red-billed Gulls can be seen at sea off the coast, along coastlines and on beaches, in estuaries and harbours, and on open grassed areas of coastal towns and cities, such as parks, sportsfields (even while games are in progress), flooded bowling greens, and school playgrounds. They occasionally venture a short distance inland, particularly along rivers, to wet pasture where they can be seen feeding, sometimes in the company of Pied Stilts and other species. They commonly associate with Black-backed Gulls. Like them, Red-billed Gulls are frequent visitors in large numbers to coastal rubbish dumps and other places where food is locally abundant, and will daily commute several kilometres between them

and their roosting sites. They are also an aggressive species. Adults attack and sometimes kill chicks of their own species that wander onto their territory in nesting colonies.

They have a very varied diet that changes somewhat with the season. During the breeding season, they feed in inshore waters mainly on planktonic crustaceans which they take close to their nesting colonies, but in autumn and winter their diet is much more varied. Over the year, it also includes small fish, all kinds of dead animal matter cast up on beaches, offal from a wide range of sources, garbage and other waste, insects and insect larvae, moths, beetles, sandhoppers, tadpoles, earthworms, lizards, scraps fed to them by humans, and even small berries of some coastal plants. Red-billed Gulls sometimes 'paddle' with their feet on wet sand or in shallow water, and on wet paddocks, presumably in order to make invertebrate food items reveal themselves. They can be seen in the breeding places of shags, gannets and terns where they pick up scraps of food. Particularly in summer, they are frequently seen not far offshore in flocks above patches of turbulent water, sometimes in the company of gannets, shearwaters and terns, where they are feeding on schools of small fish which have been forced to the surface by larger fish working from beneath them. Skuas will chase Red-billed Gulls and make them disgorge the fish they are carrying. Red-billed Gulls are themselves wellknown robbers. They have been observed to compel young shags to disgorge food, which they immediately eat. They will force oystercatchers to give up food that they have collected. They have been watched robbing Wrybills of food. If opportunity allows, some will take the eggs of their own species, and those of White-fronted and Caspian Terns which often nest with them. One showed considerable interest in the eggs of a Northern New Zealand Dotterel when the sitting adult was temporarily absent from the nest, and would no doubt have taken them if it could.

Black-billed Gull *Larus bulleri*
Description Length 35–38 cm; wingspan 81–96 cm.

Black-billed Gulls are small, slender, grey and white gulls. They are very similar in general appearance to Red-billed Gulls, but the adult of that species has a stoutish, bright red bill, bright red legs and feet, and a different upper wingtip pattern. Adult Black-billed Gulls are readily identified by their slender, longish, black bill, normally black legs and feet, and their distinctive pale-tipped upper wings in flight. The sexes are alike in plumage. Males are usually slightly larger than females, but this difference is not noticeable in the field. The entire plumage of the head, neck, underparts and tail is white. The mantle, back, and wing coverts are ash-grey. They have distinctive pale-tipped upper wings in flight because the four outer primaries are white, and the fifth and sixth are grey, all having only a narrow subterminal black bar. They have a black eye-ring which sometimes has a dark red tinge, and a white iris. Their bill is black, and normally their legs and feet are black, but some with reddish-black or even red legs can be seen in all or most months. Their gape and mouth are red. Separation of juvenile and immature Black-billed Gulls from juvenile and immature Red-billed Gulls in the field in areas where both

Adult Black-billed Gulls.

species can occur is notoriously difficult, particularly when they are at rest. Young birds of both species can look almost identical. Initially, the upper wing coverts of both are heavily spotted brownish. Those spots gradually disappear, but young Black-billed Gulls lose them sooner than young Red-billed Gulls. The amount of black in the upper wingtips of young Black-billed Gulls can vary with age. Birds of both species go through a poorly known process of change with age in the colour of their bare parts, particularly in the colouration of the bill and legs. This confusing process may be further complicated by seasonal and individual variation. There are times when the colouration of the bill and legs can be almost identical in the young of both species. Despite these difficulties, Black-billed Gulls always have a longer and thinner bill than Red-billed Gulls. The shape and length of the bill, if it can be seen well enough, may be the only reliable character by which the young of these two species can be safely distinguished in the field.

Population

Black-billed Gulls are a protected endemic species. Their total New Zealand population was recently estimated to be between 50,000 and 100,000 breeding pairs. This is consistent with the maximum number of nests counted during recent surveys. Black-billed Gulls are not as widely distributed throughout the country as are Red-billed Gulls and would appear to be less common. For example, Red-billed Gulls are common and breed in coastal North Taranaki where Black-billed Gulls are almost never seen.

Black-billed Gulls are predominantly a South Island nesting species, and Southland is clearly their major breeding stronghold. Recent surveys show that the riverbeds of Southland had more nesting Black-billed Gulls each year than the rest of New Zealand together. In the summers of 1995–96 to 1997–98, the Ornithological Society undertook a New Zealand-wide survey of nesting Black-billed Gulls. An average of about 45,500 nests each year were counted throughout the country during the two summers of 1995–96 and 1996–97. Of these, an average of about 31,500 nests (69 percent) were in Southland, followed by an average of about 12,675 nests (28 percent) in Canterbury. This survey confirmed that the number of nesting Black-billed Gulls can vary significantly from season to season, even allowing for the fact that some of the known nesting sites might not always have been counted. For example, in the summer of

Immature Black-billed Gull.

Juvenile Black-billed Gull.

1997–98 the number of nests counted in Southland fell to about 22,000, compared to the average of about 31,500 during each of the previous two summers.

Black-billed Gulls have been extending their breeding range northwards in recent decades. For a long time their most northern nesting site was the long-established inland colony at Rotorua. They were first recorded nesting at Miranda in the Firth of Thames in 1968. A breeding colony later established itself on the Karaka shellbanks in Manukau Harbour, and by 1992 a small colony had appeared in Kaipara Harbour. By the summer of 1997–98, a nesting group at Papakanui Spit was the most northern successful breeding colony of Black-billed Gulls in New Zealand.

Black-billed Gulls are less parasitic on humans than are Black-backed and Red-billed Gulls. Black-billed Gulls may not have benefited as much as those species appear to have done as a result of European settlement in New Zealand. On the one hand, arable farmland on which they feed has been created near their South Island riverbed breeding sites, and hydro-electric dams have probably lessened the chances of their nesting colonies being destroyed by flooding. On the other, the availability c⁰ suitable nesting habitat has been significantly reduced as a result of inundation by hydroelectric storage lakes, and much of the remainder has been degraded by the encroachment of introduced plants, and by its colonisation by introduced mammalian predators.

Behaviour

Black-billed Gulls are predominantly an inland species. They nest from September to February, mainly in the South Island, in large, densely packed colonies of up to 1000 pairs, on open shingle margins or islands in braided riverbeds, or on the coast on sandspits, boulderbanks or shellbanks. At that time of year they are to be found mainly inland on the larger rivers and lakes and on nearby arable land where they are often seen following the plough, frequently in the company of Black-fronted Terns. During the breeding season they feed mainly on small freshwater fish and invertebrates taken from lakes, rivers and streams, and on insects and their larvae, and other items such as grass and porina grubs taken from wet pastures and those exposed by ploughing. All birds disperse from the breeding colonies as soon as the young can fly. They disperse widely, mostly to coastal areas. A small proportion of South Island birds migrate to coastal North Island localities, mainly south of Wanganui on the west coast and south of Gisborne on the east coast. During winter they can be found frequenting estuaries, harbours, lakes, open coastlines and parks of coastal towns, frequently in the company of Red-billed Gulls. Their diet during this season is more varied than in the breeding season. It includes marine and freshwater invertebrates, small fish, scraps fed to them by humans, worms and insects. Black-billed Gulls sometimes 'paddle' with their feet on wet sand or in shallow water, presumably in order to make invertebrate food items reveal themselves. Of the three New Zealand gulls, the Black-billed Gull is the most specialised feeder and the least likely to scavenge. Skuas will chase Black-billed Gulls and make them disgorge the fish they are carrying.

Black-fronted Tern *Sterna albostriata*
Description Length 28–30 cm; wingspan 65–72 cm.

Black-fronted Terns are noticeably smaller and greyer than White-fronted Terns with which they often associate, and they always have bright orange legs and feet, irrespective of age or season. The sexes are alike. Adults have a slate grey body and wings, with a white rump and upper and under tail coverts that do not change visibly during the year. Their short, prominently forked tail contrasts with the white rump in flight. Most adult Black-fronted Terns are in full breeding plumage by late May, and all are in it from the beginning of June to early November. Birds are quite striking in this plumage. The crown from the bill to the nape and surrounding the eyes is jet black. The black crown is separated from the grey neck by a narrow, white stripe. The rump and upper and under tail coverts are white. The wings and the rest of the body are slate grey above and paler grey below. The bill, legs and feet are bright orange. Birds are in full non-

Juvenile Black-fronted Tern.

breeding plumage only from February to late April. This plumage is like the breeding plumage except that the head is pale grey, with dark patches around the eyes and on the ear coverts. The bill, legs and feet remain orange, but the tip of the bill may darken. Birds are in immature plumage for a year after losing their juvenile plumage. In immature plumage the head is grey mottled with black, and the chin is white. The patches around the eyes and on the ear coverts tend to be darker than those of non-breeding adults. The upper and lower body plumage is the same as in adults, but portions of the wings and tail are brownish. The bill colour changes progressively from very dark brown, with reddish at the base, to the orange colour of adults. The legs and feet are bright orange. Birds are in juvenile plumage for about the first three months after flying. They have a white forehead, heavily mottled crown, mantle and back, and there are black patches around the eyes and on the ear coverts. The bill is very dark brown, reddish at the base. The legs and feet are bright orange.

Population

Black-fronted Terns, the common inland tern of the South Island, are an endemic species with a population estimated to be less than 5000 breeding pairs in the early 1980s. They are migratory within New Zealand. From December to March, most birds move from their inland breeding grounds to the coast, particularly to rivermouths, estuaries, harbours and lagoons. They are present at those places from January until August when they begin returning to their inland breeding grounds. The majority remain on the east coast of the South Island, but a small proportion regularly cross Cook Strait where they can be found along the Wellington, Hawke's Bay and Bay of Plenty coasts. Some occasionally reach as far north as Kaipara Harbour, but the Bay of Plenty is the northernmost known regular wintering ground. Variable numbers of Black-fronted Terns are present there every year from late March until August. Similar numbers to those found in the Bay of Plenty also spend autumn and winter in the Napier area. The route they take to and from those localities and the South Island is not known, but sightings suggest they probably travel along the east coast of the North Island. Although Black-fronted Terns are sometimes seen at Manawatu Estuary in autumn and winter, they are very rarely recorded on the North Island west coast north of there. Autumn and winter flocks of 100–300 birds are regularly seen at some South Island localities, but smaller flocks of 20–50 are usual in the North Island.

Little is known of population changes of Black-fronted Terns. The development of farmland near their nesting sites may have created beneficial feeding opportunities. On the other hand, their nesting sites on riverbeds are susceptible to flooding, and to human disturbance, mammalian predation and trampling by stock. Many have been invaded by introduced plants such as willows, gorse, lupin and broom. Predation by introduced mammals appears to be the main cause of failure of Black-fronted Tern nests. Some young are also taken by Black-backed Gulls, and some roosting birds are taken by stoats.

Adult Black-fronted Terns flying with White-fronted Terns.

Behaviour

Black-fronted Terns breed between October and February in small colonies, often near nesting Black-billed Gulls, mainly well inland on gravel riverbeds in the South Island east of the main divide. They are fearless in defence of their nests and persistently fly noisily at intruders. During the nesting period they usually feed from fast flowing rivers and nearby farmland, but they can also be seen over lakes, tarns and farm dams. On rivers, birds always work their way upstream and then return to their starting point and work their way upstream again over the same stretch of water. They drop from a small height, without stopping, to pick up prey, usually water insects, from the surface. They will sometimes plunge-dive with open wings to catch small fish. They can frequently be seen over recently ploughed or irrigated paddocks. Feeding over land is done almost entirely on the wing, and often in the same manner as over rivers. Their main prey on farmland are earthworms and grass grub larvae, but they also take a wide variety of insects, particularly beetles. They will obtain food by following the plough, often in the company of Black-billed Gulls. During the winter, they commonly feed alone at sea, usually not far offshore. At this time of year they can also be seen feeding in groups over the lower reaches of rivers, over coastal fields with short vegetation, and

Adult Black-fronted terns.

at lagoons and coastal oxidation ponds. In coastal fields they often feed in the company of Red-billed Gulls, Pied Stilts, and other birds. They can frequently be found roosting, with or near groups of White-fronted Terns and/or Black-billed Gulls, at rivermouths or on beaches, or on the seaward shores of coastal lakes. They will also roost in fields and on posts, particularly those on tidal mudflats.

Caspian Tern *Sterna caspia*
Description Length 47–54 cm; wingspan 130–145 cm.

Caspian Terns are the largest terns in New Zealand. They are of striking and impressive appearance. Their large size and red bill make them easy to identify. Adult birds have pale grey upperparts and white underparts. Their forehead, cap and nape are black in breeding plumage, which many birds may assume by mid-July. They have a large, bright red bill, black towards the tip which is yellow, and black legs and feet. Their underwings are dark at the tips, and they have a short, slightly forked tail. In non-breeding plumage, the cap is heavily flecked with white, or may be almost white. The colours of the bill, legs and feet are the same as those of adults in breeding plumage. Immature birds are similar to non-breeding adults. The markings on the head of birds in juvenile plumage are more extensive and darker brown than in immatures, their upperparts are mottled brown with buff-white edges to the feathers, and their bill, legs and feet are dull orange. The bill progressively reddens and the legs and feet progressively darken with age.

Population
Caspian Terns are a relatively common native species that is almost cosmopolitan in its distribution, breeding locally in temperate parts of all continents except South America. In New Zealand, they breed between September and January in widely separated localities on both main islands. Some nest inland, for example, in the South Island as isolated pairs on shingle riverbeds in association with colonies of Black-backed Gulls. However, Caspian Terns are mainly a coastal nester, the majority breeding in long-established colonies on exposed shellbanks or sandbanks immediately above the high-tide level. Caspian Tern colonies often

vary in location and size from year to year. A colony normally has from 5 to 20 pairs, but some colonies can be considerably larger. For example, one on Matakana Island near Tauranga had 148 pairs in 1992, and another on Rat Island in Kaipara Harbour had about 130 nests in 1997. Caspian Terns that nest in northern New Zealand, and their offspring, do not seem to move very far when breeding has been completed. There is an autumn influx into such localities as the Firth of Thames and Manukau Harbour. It is then quite common in those places to see large numbers of Caspian Terns resting at high tide among shorebirds of several species. Caspian Terns were counted in the Firth of Thames and Manukau Harbour during summer and winter over a 20-year period from 1962 to 1981. Those counts show that in the Firth of Thames three times as many Caspian Terns can be present in winter as in summer, and on Manukau Harbour there can be twice as many. On the other hand, some South Island birds and their offspring move a considerable distance north after breeding. In autumn, birds from a colony near Invercargill stop temporarily at such places as the Avon-Heathcote and Manawatu estuaries.

The New Zealand population of Caspian Terns is probably less than 3500 birds. Caspian Tern nesting colonies are noisy and conspicuous. Many are subject to mammalian predation and human disturbance. Black-backed Gulls can be major predators of eggs and small chicks. Red-billed Gulls predate their eggs. Lake Ellesmere may be the only Canterbury coastal locality where successful colonial nesting by Caspian Terns has not been prevented by habitat modification and disturbance. It

is not known what effect those factors, among others, may be having on the overall number of Caspian Terns in New Zealand.

Behaviour

Caspian Terns are primarily birds of shallow coastal waters, most commonly frequenting coastlines, harbours and estuaries, coastal lakes, and the lower reaches of rivers. They are seldom seen far out at sea. Nor are they often seen far inland, but they do regularly frequent freshwater lakes on the Volcanic Plateau and penetrate well up the lower Waikato River. The flight of Caspian Terns is rather more direct, purposeful and less buoyant than in most terns, but with its

Juvenile Caspian Tern (right) begging from adult.

steady, shallow wingbeats it is still graceful, easy and unmistakably tern-like. They have a harsh grating call, quite frequently and loudly emitted, particularly when in flight. They are usually solitary feeders. Caspian Terns feed mainly on surface swimming fish. They hunt in a characteristic manner, usually flying about 5–10 metres above the water with their bill pointed downwards. When they sight suitable prey they hover for a few seconds then plunge steeply into the water. They often completely submerge themselves briefly and, if successful, will sit on the surface for a short while before again taking flight. Juveniles accompany their parents to wintering localities, and can frequently be seen persistently begging food from them. Begging typically lasts for 10–30 seconds and is accompanied by high-pitched whining calls.

White-fronted Tern *Sterna striata*
Description Length 35–43 cm; wingspan 79–82 cm.

White-fronted Terns are medium-sized whitish terns, with a white fore-head of variable width which separates a black cap and bill. The sexes are alike, except that females are usually slightly smaller but this difference is not noticeable in the field. Adult White-fronted Terns have pale grey upperparts and white underparts with a black crown and nape, variable white forehead and deeply forked white tail, the depth of the fork depend-ing on the season. The upper wing is entirely pale grey except for a black outer web to the outermost primary feather. The bill of White-fronted Terns is black with a very pale tip, and the legs are black or reddish black irrespective of age or season. In breeding plumage, which a considerable

proportion of adults have assumed by at least early August, they have a jet black crown and nape, with a nar-row white forehead and white lores. Many have a conspicuous pinkish tinge on their breasts at this time. They have long outer tail streamers that are obvious in flight and extend well beyond the wings when the bird is at rest. In non-breed-ing plumage, the cap recedes to above the eyes, creating a much broader white forehead, and the middle of the crown is mottled. Immature birds are similar to non-breeding adults, but their middle and lesser upper wing coverts sometimes have traces of light brown mottling which shows as a black line on the inner upper forewing. Their outer primary feathers are dark in colour, so the outer edges of their upper wings appear darker in flight than those of adults. Birds in juvenile plumage are very distinc-tive. They have dark patches before and behind the eyes. Their crown and nape is streaked with black, white and buff. Their upperparts are barred and mottled ash grey. Dark mottling on the inner upper forewing, and their dark outer primary feathers, are prominent both when they are at rest and in flight. Their underparts are white.

136

Population

White-fronted Terns are by far the most common tern around the New Zealand coast, and are often seen in large flocks. Nevertheless, little is known about their distribution in New Zealand at any particular time of the year. Every autumn after breeding a considerable number of young birds and some adults migrate to the south-eastern coast of Australia where they are widespread between May and November. However, many adults remain on New Zealand coasts throughout the winter.

The nesting colonies of White-fronted Terns, particularly those on sandy beaches and on shingle or shellbanks, are subject to destruction by storm and human interference. The encroachment of introduced plants such as lupin significantly reduced their nesting habitat on some braided riverbeds of the eastern South Island. Breeding adults, and eggs and chicks, often suffer severe predation, especially by mustelids, cats and rats. Stoats will take their eggs and chicks. Ship rats prevented them nesting on an offshore islet near New Plymouth. If opportunity allows, Black-backed and Red-billed Gulls will take their eggs and chicks. White-fronted Terns among large groups roosting on open beaches are sometimes predated by cats. Nevertheless, White-fronted Terns are still common, with a population estimated to be in excess of 100,000 breeding pairs in the early 1980s, but they may not be as plentiful as they were. Little information is available on their population trends. They are difficult to monitor because they often do not use the same nest site in successive years, and their breeding success is highly variable from season to season.

Adult White-fronted Tern.

Juvenile White-fronted Tern.

Surveys of nesting colonies of White-fronted Terns were carried out throughout New Zealand by members of the Ornithological Society during the breeding seasons of 1995–96 to 1997–98. A considerably greater number of nests was counted in 1997–98 than in either of the previous two breeding seasons. This difference can at least be partly explained in the case of the 1995–96 breeding season when many colonies were destroyed by storms or flooding due to unusually adverse weather conditions. The surveys show that in the three seasons counted most White-fronted Terns nested in Otago (particularly at the Waitaki River), followed by Auckland (particularly at Papakanui Spit in Kaipara Harbour), and Northland (particularly at Walker Island in Rangaunu Harbour).

Behaviour

White-fronted Terns nest mostly on the coast throughout New Zealand between October and February. They mainly nest in larger colonies of varying sizes, sometimes numbering many hundreds or even thousands of pairs, on sandy beaches and on shingle or shellbanks. Smaller colonies

and isolated pairs nest on the shores of coastal lakes, and on offshore rock-stacks and coastal cliffs. A quite sizeable colony nested unsuccessfully among large boulders in a harbour reclamation at New Plymouth. They often nest with or near other species, principally Red-billed and Black-billed Gulls, and Caspian Terns. White-fronted Terns will sometimes, for no apparent reason, abandon a nesting site that they have successfully used and not use it again until two or more years later.

White-fronted Terns, which are buoyant and graceful on the wing, feed almost exclusively in coastal waters, up to a few kilometres out to sea, but at times some may search for food in tidal creeks and rivers. They are rarely seen inland. Their food is small fishes, especially those that congregate near the surface in shoals, which they catch by shallow plunge-diving. Particularly in summer, they are frequently seen not far offshore in large flocks above patches of turbulent water, sometimes in the company of gannets, shearwaters and gulls, where they are feeding on schools of small fish that have been forced to the surface by larger fish working from beneath them. The terns fly into the wind not far above the surface of the water, and dart down and pick up their prey.

Early in the mating season, when many birds are in breeding plumage, there is often constant coming and going from roost sites with birds moving to fish offshore and bringing the small fish they have caught back and offering it on the ground to other birds. This process is accompanied by much displaying and considerable clamour. In summer, especially near nesting colonies, they are likely to be attacked in flight by skuas that harass them until they drop a fish they are carrying, which the attacker then catches before it reaches the water. Red-billed and Black-billed Gulls will also attempt to rob them of their fish, but usually the terns are more dexterous than the gulls and escape their pursuers. Sometimes considerable numbers of White-fronted Terns will roost at such places as the mouths of streams, open beaches, sand and shell-banks, the margins of coastal lakes, and rock and concrete walls in harbours. They will also roost on jetties and posts or piles in water. White-fronted Terns are very fond of bathing in both salt and fresh water.

Previous page: Eastern Broad-billed Sandpiper at Manawatu Estuary.

New Zealand waders

Black Stilts are probably the world's rarest wading birds with only 37 known adults in the wild in 1999. The species has been under intensive management for many years. The few remaining birds breed naturally only in the Mackenzie Basin of inland South Canterbury. Black Stilts sometimes interbreed with Pied Stilts. The Black Stilts mated with Pied Stilts, and the hybrid birds which result from those matings, tend to migrate to the coast and northwards at the end of the breeding season. Those birds are occasionally seen at some of the outstanding and significant shorebird localities.

Shore Plovers are an endangered endemic species that now breed naturally only in the Chatham Islands, where the total population fluctuates between about 110 and 150 birds. In recent years some Shore Plovers bred in captivity have been released on Motuora Island in Hauraki Gulf and on Portland Island in Hawke Bay. Individuals are occasionally seen on nearby estuaries.

Black-fronted Dotterels, which breed throughout Australia, began to colonise New Zealand in the late 1950s. They now breed on shingle riverbeds of the eastern and southern North Island, and the north-eastern and eastern South Island. Most of the estimated New Zealand population of 1700 birds stay on those rivers after the breeding season, but some form small flocks at favoured lagoons, lakes, estuaries and sewage ponds. Black-fronted Dotterels are occasionally seen at some of the outstanding and significant shorebird localities.

Three other species of wader will not be seen on the New Zealand mainland. **Chatham Island Oystercatchers** and **Chatham Island Snipes** are restricted to the Chatham Islands. **New Zealand Snipes** are now restricted to the Snares, Antipodes and Auckland islands.

Uncommon visiting and vagrant waders

Uncommon visiting waders are those species that are present in New Zealand every year, or in most years, but in very small numbers. Included in this group are **Large Sand Dotterels, Mongolian Dotterels, Grey Plovers, Sanderlings, Pectoral Sandpipers, Little Whimbrels, Asiatic Black-tailed Godwits, Hudsonian Godwits, Wandering Tattlers, Grey-tailed Tattlers, Greenshanks, Marsh Sandpipers,** and **Terek Sandpipers.**

Vagrant waders are those species that have been recorded in New Zealand on only one occasion, or on not many occasions, over the years. Included in this group are **Painted Snipes, Australian Red-necked Avocets, Oriental Pratincoles, Ringed Plovers, Oriental Dotterels, Red-kneed Dotterels, American Golden Plovers, Japanese Snipes, Great Knots, Dunlins, Baird's Sandpipers, White-rumped Sandpipers, Little Stints, Long-toed Stints, Western Sandpipers, Eastern Broad-billed Sandpipers, Stilt Sandpipers, Ruffs, Asiatic Dowitchers, Upland Sandpipers, Common Sandpipers, Lesser Yellowlegs, Grey Phalaropes, Red-necked Phalaropes,** and **Wilson's Phalaropes.**

Uncommon waders, and particularly vagrants, account for a very small proportion of the total number of waders annually visiting New Zealand, but they attract a disproportionate amount of attention from birdwatchers.

Grey-tailed Tattler with Bar-tailed Godwit at Manawatu Estuary.

Terek Sandpiper at Manawatu Estuary.

Skuas

Brown Skuas, which breed in the subantarctic and parts of southern New Zealand, are occasionally seen on the New Zealand mainland in winter. **South Polar Skuas**, which breed in Antarctica, are sometimes beach-wrecked or seen off the New Zealand coasts. Two skuas which breed in arctic and subarctic regions, **Arctic Skuas** and **Pomarine Skuas**, are regular visitors to New Zealand waters from spring to autumn. **Long-tailed Skuas**, another arctic and subarctic breeding species, occasionally appear in New Zealand waters from their usual wintering range in the eastern and central Pacific. Skuas are piratic. They acrobatically pursue terns, gulls, shags and shearwaters, and force them to drop or disgorge their food. From spring to autumn, Arctic Skuas, which are by far the most numerous skua seen off the coasts of the main islands of New Zealand, can often be seen harrying White-fronted Terns, and also Red-billed and Black-billed Gulls, and robbing them of their catch. Arctic Skuas sometimes come close inshore in sheltered bays and harbours, such as the Firth of Thames, and occasionally join roosting flocks of terns or gulls.

Uncommon and vagrant terns and noddies

White-winged Black Tern at Manawatu Estuary.

Fairy Terns are the rarest of New Zealand's resident birds, with a population of only about 20 individuals in 1999. They have been under intensive management for many years. Fairy Terns have a very restricted distribution in New Zealand and breed precariously at just three protected sites: at Mangawhai and Waipu on the Auckland east coast, and at Papakanui Spit in Kaipara Harbour. Nearly all the Fairy Terns that breed at Mangawhai and Waipu, and their progeny, spend the post-breeding period in Kaipara Harbour, particularly in the vicinity of Waikiri Creek.

    Eastern Little Terns are seen in New Zealand each year, either as single birds or in very small groups, at suitable localities such as Kaipara and Manukau Harbours, the Firth of Thames and Manawatu Estuary. A small number of **White-winged Black Terns** reach New Zealand each year where they might be found in a variety of plumages at some estuaries,

147

coastal lagoons and oxidation ponds. Two of their favoured localities are Ahuriri Estuary at Napier and Cooper's Lagoon in coastal Canterbury. They appear to have been present in New Zealand in unusually large numbers during the summer of 1998–99.

Eastern Common Terns and **Arctic Terns** are probably annual visitors to New Zealand but can be difficult to separate from the White-fronted Terns with which they often associate. **Antarctic Terns** breed on some islands off Stewart Island, but they have not yet been recorded north of Foveaux Strait. **Gull-billed Terns** have been recorded here on a number of occasions. **Whiskered Terns** and **Crested Terns** are sometimes seen in New Zealand. **Sooty Terns, White Terns and White-capped Noddies** occasionally reach this country, most often after northerly gales. Small numbers of **Grey Ternlets** have bred on islands off the north-eastern North Island and are occasionally seen in northern coastal waters. **Common Noddies** have been very infrequently recorded, and so far only one **Bridled Tern,** a beach-wrecked specimen, has been recorded in New Zealand.

An observer who wishes to specifically identify any of the birds mentioned in this section should first consult *The Field Guide to the Birds of New Zealand* (Heather and Robertson, 1996) and/or *The Hand Guide to the Birds of New Zealand* (Robertson and Heather, 1999).

Glossary

Adult a bird that has reached its fullest development.

Breeding plumage the plumage worn by a bird during the nesting period.

Ear coverts the small feathers that cover the region of a bird's ears.

Endemic New Zealand bird a species whose natural breeding range is in New Zealand and nowhere else.

Eye ring or eye lid the bare skin around a bird's eye.

Front the forehead between the base of the upper mandible and the crown.

Gape fleshy skin at the sides of the base of the bill.

Immature stage of plumage between the first moult and full breeding plumage (= subadult).

Iris the thin tissue in front of the lens of the eye.

Juvenile birds in their first plumage after replacing natal down.

Lores the area between the base of the upper mandible and the eyes.

Mandibles the upper and lower parts of a bird's bill.

Migrant a species that moves annually and seasonally between breeding and non-breeding areas.

Nape the back of a bird's neck.

Native New Zealand bird a species that is naturally found in New Zealand, including recently self-introduced species.

Non-breeding plumage the plumage worn by a bird outside the nesting period.

Phase a regularly occuring colour variant.

Primary feathers the flight feathers of the outer wing.

Pupil the opening in the iris of the eye.

Secondary feathers the flight feathers of the inner wing.

Superciliary streak or stripe a marking above the eye of a bird.

Vagrant a species that occurs in a given area very infrequently, and whose normal range is in another area.

Wattles   colourful fleshy drupes on either side of the gape.

Wing bar   a band of contrasting colour on the wing, usually formed by the tips of the wing coverts.

Wing coverts   feathers that cover the bases of the main wing feathers.

Bibliography

Chambers, S. 1989. *Birds of New Zealand: Locality Guide.* Arun Books: Hamilton.

Cromarty, P. and Scott, D.A. 1996. *A Directory of Wetlands in New Zealand.* New Zealand Department of Conservation.

Harrison, P. 1985. *Seabirds: An Identification Guide.* Christopher Helm: London.

Hayman, P., Marchant, J. and Prater, T. 1986. *Shorebirds: An Identification Guide to the Waders of the World.* Croom Helm: Beckenham.

Heather, B.D. and Robertson, H.A. 1996. *The Field Guide to the Birds of New Zealand.* Viking: Auckland.

Higgins, P.J. and Davies, S.J.J.F. (Eds). 1996. *Handbook of Australian, New Zealand and Antarctic Birds. Vol. 3.* Oxford University Press: Melbourne.

Marchant, S. and Higgins, P. (Eds). 1993. *Handbook of Australian, New Zealand and Antarctic Birds. Vol. 2.* Oxford University Press: Melbourne.

O'Donnell, C.F.J. 1985. 'Lake Ellesmere: A Wildlife Habitat of International Importance', *Fauna Survey Unit Report,* No. 70. NZ Wildlife Service.

Oliver, W.R.B. 1955. *New Zealand Birds.* Second edition. A.H. & A.W. Reed: Wellington.

Owen, S.J. (Ed.). 1992. *The Estuary. Where our Rivers Meet the Sea.* Christchurch City Council.

Petyt, C. 1999. *Farewell Spit. A Changing Landscape.* Terracottage Books: Takaka.

Robertson, C.J.R. (Ed.). 1985. *Reader's Digest Complete Book of New Zealand Birds.* Reader's Digest: Sydney.

Robertson, H.A. (Ed.). 1999. 'Wader Studies in New Zealand. A Tribute to Richard B. Sibson (1911–1994) and Barrie D. Heather (1931–1995), *Notornis* 46(1): 1–242.

Robertson, H.A., and Heather, B.D., 1999. *The Hand Guide to the Birds of New Zealand.* Penguin Books: Auckland.

Turbott, E.G. (Conv.). 1990. *Checklist of the Birds of New Zealand.* Third edition. Random Century and Ornithological Society of New Zealand: Auckland and Wellington.

Index of common names

Index of scientific names